How
Ronald Reagan
Changed My Life

ALSO BY PETER ROBINSON

It's My Party: A Republican's Messy Love Affair with the GOP

Snapshots from Hell: The Making of an MBA

—— HOW ——
RONALD REAGAN
CHANGED MY LIFE

PETER ROBINSON

ReganBooks
An Imprint of HarperCollins*Publishers*

HarperCollins books may be purchased for educational, business, or sales
promotional use. For information please write: Special Markets Department,
HarperCollins Publishers Inc., 10 East 53rd Street, New York, NY 10022.

FIRST EDITION

Designed by Judith Stagnitto Abbate/Abbate Design

Printed on acid-free paper

Library of Congress Cataloging-in-Publication Data
Robinson, Peter, 1957–
How Ronald Reagan changed my life / Peter Robinson.— 1st ed.
p. cm.
ISBN 0-06-052399-9
1. Reagan, Ronald—Influence. 2. Reagan, Ronald—Friends and associates.
3. Presidents—United States—Biography. 4. Robinson, Peter, 1957–
5. Speechwriters—United States—Biography. I. Title.

E877.2.R63 2003
973.927'092—dc21
[B]   2003046777

03 04 05 06 07 BVG/RRD 10 9 8 7 6 5 4 3 2 1

CONTENTS

PROLOGUE

JOURNAL ENTRY, NOVEMBER 1999:

Last night during the taping of a Fox television program, The Real
Reagan, *the host, Tony Snow, asked each member of the panel to sum up
Reagan's place in history. I found myself launching into a little perora-
tion. "Ronald Reagan's beliefs were as simple, unchanging, and
American as the flat plain of the Midwest where he grew up. He placed
his faith in a loving God, in the goodness of the country, and in the wisdom of the people. He applied those beliefs to the great challenges of his
day. In doing so he became the largest and most magnificent American of
the second half of the twentieth century."*

*If any of my friends see the program, I suppose they'll take it for
granted that I was overstating the case for the cameras. They've certainly
never heard me talk that way over lunch on the Stanford campus or at
dinner parties in Palo Alto, where we always lace our conversation with
a knowing dose of cynicism.*

The odd thing is, I meant every word.

THIS BOOK IS not a memoir or exposé. It's a primer. It presents ten precepts—ten life lessons—that I learned from Ronald Reagan.

Although I learned these lessons more than a dozen years ago, I still think of them every day. The lessons are concrete—maxims I find *useful.* And because I drew them from the life of a near-contemporary, they exercise a particular claim on me. Washington, Adams, Jefferson, Lincoln—each is so remote that he might as well have been a figure of legend. But Reagan knew the America that I know. He saw what I see. If he lived up to these precepts, then I ought to be able to do so myself.

The reader is entitled to wonder what makes me think I'm qualified to write this book. When I worked for Reagan, serving six years in the White House, I was never a member of his senior staff, holding only a junior position instead. Just how junior I learned one day as I climbed into a government car.

The car sat at the end of a long line of vehicles on the drive that sweeps across the South Lawn. In formation for a motorcade from the White House to a hotel where the President would be delivering a speech, the vehicles idled as members of the White House staff boarded them, each member of the staff entering the vehicle to which he had been assigned, high-ranking members of the staff climbing into cars near the front, middle-ranking members into cars in the middle, and so on, all the way down to the last car and me. Next to the President's limousine, which rumbled at the head of the line, stood Michael Deaver, the President's deputy chief of staff. As Deaver waited for the President to emerge from the Oval Office, he cast his gaze down the line of vehicles. It came to rest on my car. Realizing I had the car to myself—by now each of the others had been boarded by two or three members of the staff—Deaver waved to one of his assistants, pointed to me, and said a few words. Deaver's assistant strode over. Then he kicked me out of the motorcade, instructing my driver to take the car back to the garage. Deaver wasn't about to spend taxpayer money transporting anyone as lowly as me.

Even though I was the kind of person who found himself being told to climb out of government cars almost as soon as I had climbed in, I possess the only two qualifications for writing this book that I need. They're peculiar qualifications, I'll admit. But they'll do. The first is that I was a Reagan speechwriter. The second is that when I took the job, going to work in the Reagan White House when I was just twenty-five, I was callow and unformed.

As a speechwriter, I became an expert on Reagan. All his speechwriters did. We had to. It was the only way we could do our work. We mastered every position he had ever taken, poring over his old speeches, radio talks, and newspaper columns. We scribbled down every word he uttered when we met with him in the Oval Office. And we watched him work. It was an unwritten rule in the speechwriting shop that whenever the President delivered a speech that you yourself had written, you yourself showed up to observe. (When Michael Deaver kicked me out of the motorcade, forcing me to break this rule, the speechwriting office protested. I doubt Deaver paid much attention to the protest. On the other hand, I can't recall that a speechwriter was ever kicked out of a motorcade again.)

Since leaving the White House I've held jobs in a number of organizations, including the News Corporation, a company in which every big decision and a startling number of small ones were made by the man who built it, Rupert Murdoch. Yet nowhere have I encountered a preoccupation with one person as intense as the preoccupation of the White House speechwriters with President Reagan. Maybe the Swiss Guard demonstrates a similar preoccupation with the Pope, but I have my doubts. The job of the Swiss Guard is only to protect the pontiff. The speechwriters were trying to inhabit Reagan's mind.

Probably the best evidence of our absorption with the President lay in the Reagan impersonation that each of us developed. Since it is perfectly possible for words to look just right on paper but sound all wrong spoken aloud, when a speechwriter finished a draft he would test it by reading passages as if he were the President himself. A rough job of mimicking Reagan's phrasing and intonation would have done the trick. But we all identified with Reagan so completely that when the speechwriter opened his mouth it would be the President's voice that came out. If you'd walked past our offices in the Old Executive Office Building one evening when several of us faced deadlines, you'd have thought you'd wandered into a lunatic asylum in which the inmates believed they were not Caesar or Napoleon, but Reagan.

Lots of people found our impersonations hilarious—even today, my Reagan never fails as a party trick. Once, however, I performed Reagan for a Secret Service agent. He looked at me stonily for a moment, then said, "There's a law against impersonating federal officials." That shut me up.

This brings me to my second qualification. It is, in a sense, an anti-qualification, not something I possessed but something I lacked: in a word, my immaturity. I was the youngest member of the speechwriting staff. I was also the most inexperienced. I'd come from a small town in upstate New York. Writing speeches was the first full-time job I'd ever held. Aware that I had some growing up to do, I found myself looking around for a model, a wise, successful adult from whom I could learn. My college friends, who had joined corporations, banks, and law firms, had already found their models. They were studying the savviest managers in their corporations, the richest directors in their banks, and the suavest partners in their law firms. Whom could I study? It made no sense to model myself on the other speechwriters. Most of them were only in their thirties. For a while I thought myself unlucky. Then I realized I might as well have won the twentysomething lottery. My model could be Ronald Reagan.

I began studying the President not just to write speeches for him, but to figure out how he did it. That question—How did Reagan do it?—is worth dwelling upon. I found myself asking it in one form or another until the day I left the White House. I still remember where I was and how I felt the first time I found myself asking it, which was the first time I saw the President in person.

A couple of weeks after joining the Office of the Vice President, where I worked for a year and a half before joining the President's staff, I huddled late one evening with Chase Untermeyer, an adviser to Vice President Bush, in Chase's tiny office in the West Wing. Exhausted—we'd put in a long day—Chase and I were struggling to rewrite a speech I'd drafted. Out of the corners of our eyes we saw a group of men striding down the hallway toward us. Somewhere in the background a vacuum cleaner was droning away, and I suppose if at that very instant you'd asked us to identify the men in the hallway we'd have replied that they were part of the janitorial staff. Suddenly we realized that two members of the group were Secret Service agents, while the third, walking between them, was the President of the United States.

Just outside Chase's office, the hallway down which the

President and the agents were walking joined a second hallway, forming an L. When the lead agent, who was a few steps ahead of the President, reached the corner of the L, he performed a neat right-angle turn, entering the second hallway. A moment later, the President and the other agent began turning into the second hallway themselves. Then the President spotted Chase and me. In a single, graceful movement, he adjusted direction, stepping to the door of Chase's office. He put his head in the door, grinned, waved, and gave us a wink. Before Chase or I could stand up, much less say anything, the President stepped back into the hallway and disappeared, taking the second agent with him.

That little encounter with the President won't show up in any history book, needless to say. It didn't even make for much of a story when I telephoned my parents that night—they couldn't understand why I hadn't shaken the President's hand and told him who I was. But twenty years later, I can still replay the encounter in my mind as if in slow motion. The grin. The wave. The wink. Afterward, Chase and I felt better. We felt refreshed. The sheer sweetness of the man had washed over us like a bath.

"How," I asked, turning to Chase, "did such a nice guy ever get to be President?"

How indeed. Politicians ordinarily demonstrate at least a rough congruence or fit between their personalities and their careers. Lyndon Johnson and Richard Nixon were crude, vulgar, aggressive, and suspicious. If politics were a greasy pole, you could imagine just how they had hauled themselves to the top. Even John Kennedy and Jimmy Carter possessed a kind of suppressed intensity that made it possible to understand how they had achieved their political success. Ronald Reagan? He seemed altogether too genial and relaxed. When Jack Warner, Reagan's old boss at Warner Brothers, was told that Reagan was running for governor of California, Warner misunderstood. "No," Warner replied, thinking the conversation was about a new movie, not a political campaign, "Jimmy Stewart for governor. Ronald Reagan for best friend."

How did Reagan do it? When I found myself wondering how such a nice guy ever got to be President, I had merely encountered the question in one of its many guises. How did a former movie actor persuade the American people to take him seriously? How did someone who graduated from an obscure college in the Midwest where by his own admission he paid little attention to his studies, someone former secretary of defense Clark Clifford famously termed an "amiable dunce," develop the most sweeping policy agenda since the New Deal? How did such a relaxed, genial man confront the Soviet Union so ruthlessly?

For that matter, how did a chief executive in his seventies stand up to the simple physical demands of the office? True, Reagan was known to nod off during meetings, just as the press so gleefully reported. Yet for every instance in which the President dozed for a moment or two in the Cabinet Room, everyone who worked around him could name five instances in which he demonstrated remarkable stamina. (They could also name a couple of instances in which they'd nodded off themselves. Just try listening to a member of the Council of Economic Advisers drone on about bond rates.)

One year we kept the President waiting for the final text of the State of the Union Address until eight o'clock in the evening, just an hour before he had to deliver it. I remember imagining my state of mind if I'd been waiting for the speech myself. That was easy. I'd have been having a panic attack—sweaty palms, dry mouth, heart palpitations. By the time I received the text, I'd have been in no condition to deliver it. But the President calmly reviewed the text in his limousine during the ride to Capitol Hill, then went before both houses of Congress and tens of millions of television viewers to deliver the forty-five-minute address flawlessly. The next morning he arrived at the Oval Office looking just the way he always did, which is to say as fresh and at ease as if he had just put in a couple of hours at a health club. How did Reagan do it?

When I first started asking the question I wasn't sure I'd ever get any answers. I couldn't get a purchase on Reagan. He seemed too much larger than life. There were days when I felt I might as well have chosen to model myself on one of the faces on Mount Rushmore. Then, although Reagan never ceased to elicit my awe, he gradually began assuming the dimensions of an ordinary mortal. Eventually I came to see him not only as a former movie star, as the leader of the free world, and as the Great Communicator, but as a man. An incident comes to mind here. When in 1985 the President returned from the hospital after undergoing cancer surgery, the then–chief of staff, Donald Regan, arranged for him to address the entire White House staff in the East Room. Regan wanted us to see that the President was back to normal. It almost worked. The President spoke just as well as he always did. He even had the old twinkle in his eye. But he had lost so much weight that his shirt collar was two sizes too big. If you ever need proof that someone who seems larger than life is human after all, look at the gap between his neck and his collar after he's had a bout with cancer.

Once I understood that he had no more good health, stamina, or time than any other person, I began to see how Reagan did it. I realized, for instance, that although the President made a show of taking a relaxed approach to his job—"I know hard work never killed anyone," he once quipped, "but I figure why take a chance?"—he actually worked steadily. Whereas Vice President Bush often wouldn't get around to looking at a speech until he was actually delivering it, the President edited every draft the speech-writers sent to him, condensing material, enlivening flat passages, and firming up arguments. Reagan's diligence informs chapter two of this book. I learned the extent to which the President relied on his wife, drawing on her for comfort and encouragement, the observation to which chapter seven is devoted. I saw one incident after another in which Reagan demonstrated the gentleness, courtesy, and good humor that inform chapter nine.

Hard work. A good marriage. A certain lightness of touch. The longer I studied Ronald Reagan, the more lessons I learned.

OH, ALL RIGHT. I'll come clean. I landed in the White House on a fluke.

In the spring of 1982, I found myself living in a dank cottage on the outskirts of Oxford, England. I'd attended Oxford for two years after graduating from Dartmouth, then remained in England an additional year to write a novel. Or rather, half a novel. My material proved so bad that even I saw no point in completing it. Broke, I sent letters to everyone I thought might be able to give me a lead on a job. William F. Buckley Jr. was among those who replied. This was gracious of him. The only claim I could make on Buckley's attention was that he'd complimented a couple of pieces I'd published in the *Daily Dartmouth*. Buckley suggested that I get in touch with his son, Christopher, who, as chief speechwriter for Vice President Bush, might know of a few speechwriting jobs in Washington. This was more than gracious of him. It was the only lead I received. At the beginning of July I packed up my cottage, took a bus to Heathrow Airport, and flew to Washington. Two days later I presented myself at the office of Christopher Buckley.

What did I expect? At best, I thought, Christopher might offer me a few polite words of advice, then make some telephone calls on my behalf—I distinctly recall concluding that I'd be lucky if Christopher helped me land a job with the postmaster general. Instead Christopher surprised me. He was planning to leave his own job in a couple of weeks, he explained, and since his replacement, who had been lined up for months, had just pulled out, I might as well replace him myself. Me? Write for the Vice President? Now the author of five comic novels, a collection of satirical essays, and other works of humor, Christopher had probably detected amusing possibilities the moment I walked through his door. He took a copy of my résumé, promised to recommend me to the Vice President's press secretary, then suggested I go downstairs to meet the President's chief speechwriter, Tony Dolan, who might know of a few speechwriting jobs himself.

As I was introducing myself to Tony five minutes later, his telephone rang. "You need a speechwriter?" Tony said into the telephone, shoving the unlit cigar he'd been chomping to one corner of his mouth. Covering the mouthpiece, Tony whispered that the person on the other end was the campaign manager for Lew Lehrman, the businessman then running against Mario Cuomo for governor of New York. "It so happens I have a speechwriter standing right here," Tony said, speaking into the telephone once again. "No, really. Christopher Buckley just sent him down." When he hung up, Tony saw no more reason why I shouldn't write speeches for a gubernatorial candidate than Christopher had seen why I shouldn't write speeches for the Vice President.

That afternoon, Christopher and Tony conspired, and then,

the following day, they put their plan into effect. Christopher told the Vice President's press secretary that he'd found a new replacement, namely me, but that he'd better hire me quickly because I was being wooed by the Lehrman campaign. Tony told the Lehrman campaign that I was just the speechwriter they needed, but that they'd better hire me quickly because I was being wooed by the Vice President's staff. You couldn't accuse Christopher or Tony of telling whoppers. Not exactly, anyway. As soon as they'd spun them, their tales became true. Both the Vice President's staff and the Lehrman campaign *did* begin wooing me. Over the next two weeks the Lehrman campaign flew me to New York three times while the Vice President's press secretary had me return to the Old Executive Office Building for a series of interviews.

Each convinced that the other was about to hire me, neither the Bush staff nor the Lehrman campaign made any effort to investigate my qualifications—as best I can recall, nobody even bothered to ask for a writing sample. When the Bush staff and the Lehrman campaign both made offers, I decided to go to work for the Vice President. Whereas Lehrman might never become governor, I reasoned, Bush would remain Vice President for at least two more years.* Noble motives? I wanted job security.

You see my point, of course. I got a job as a White House speechwriter without ever having written a speech in my life. And that wasn't all. When I left the office of the Vice President to join the President's speechwriting shop a year and a half later, I was able to do so only because two of the President's speechwriters had quit at the same time, forcing Bently Elliott, the director of presidential speechwriting, to scrounge for replacements. He hired me because I was already in the building.

---

*Lehrman did indeed lose to Cuomo, although by less than 3 percent.

Which brings us back to my qualifications for writing this book. Serving for six years as a speechwriter gave me the opportunity to conduct a close, continuous study of one of the largest and most magnificent Americans in the history of the republic. Landing in the White House on a fluke made me conscious that I'd better do just that.

IN WORKING on this book, I conducted several dozen interviews, refreshing my memory by talking with my old colleagues in the speechwriting shop, with other members of the White House staff, and with Ronald Reagan's sons, Michael and Ron. Dialogue from these interviews, which I've edited only for grammar and clarity, appears for the most part in the present tense. Dialogue set during the 1980s appears, by contrast, in the past tense. Some of this dialogue, including many of the comments by the President himself, is based on notes I made at the time, scribbling either on a legal pad during office hours or in my journal after I'd gone home. But some is based on my memory alone. If Ronald Reagan had been like Lyndon Johnson or Richard Nixon, I'd have been able to check a lot of this dialogue against the tape recordings he'd have had made. But if Ronald Reagan had been like Lyndon Johnson or Richard Nixon, he wouldn't have been Ronald Reagan.

Although it took me a while to sit down and start writing, I can tell you precisely when I conceived of this volume: while watching the evening news of February 6, 2001. My oldest child, then nine and a half, happened to wander into the room during a segment marking Ronald Reagan's ninetieth birthday. She watched for a moment. Then she turned to me and asked, "Dad, is that the President you worked for?"

What answer could I give her? How could I make her see? I wanted my daughter to recognize that the world she inhabited was freer and more prosperous because of that old, old man on television. But I also wanted her to grasp my personal debt to him, enabling her to understand all that he taught me—how to work and how to relax, how to think and how to use words, how to be a good husband, how to approach life itself. Although my parents, her grandparents, taught me more than anyone else, I found myself wanting to explain, I encountered President Reagan at that impressionable moment when I had stepped out on my own for the first time. As I worked for him, he shaped me.

My experience in the White House had always been so much a part of me that I'd taken it for granted. Now I saw that for the sake of my daughter, and, later, when they grew old enough to understand, her siblings, I needed to put it down.

I needed to tell my children how Ronald Reagan changed my life.

One

THE PONY
IN THE
DUNG HEAP

When Life Buries You, Dig

JOURNAL ENTRY, JUNE 2002:

Over lunch today I asked Ed Meese about one of Reagan's favorite jokes. "The pony joke?" Meese replied. "Sure I remember it. If I heard him tell it once, I heard him tell it a thousand times."

The joke concerns twin boys of five or six. Worried that the boys had developed extreme personalities—one was a total pessimist, the other a total optimist—their parents took them to a psychiatrist.

First the psychiatrist treated the pessimist. Trying to brighten his outlook, the psychiatrist took him to a room piled to the ceiling with brand-new toys. But instead of yelping with delight, the little boy burst into tears. "What's the matter?" the psychiatrist asked, baffled. "Don't you want to play with any of the toys?" "Yes," the little boy bawled, "but if I did I'd only break them."

Next the psychiatrist treated the optimist. Trying to dampen his out-

look, the psychiatrist took him to a room piled to the ceiling with horse manure. But instead of wrinkling his nose in disgust, the optimist emitted just the yelp of delight the psychiatrist had been hoping to hear from his brother, the pessimist. Then he clambered to the top of the pile, dropped to his knees, and began gleefully digging out scoop after scoop with his bare hands. "What do you think you're doing?" the psychiatrist asked, just as baffled by the optimist as he had been by the pessimist. "With all this manure," the little boy replied, beaming, "there must be a pony in here somewhere!"

"Reagan told the joke so often," Meese said, chuckling, "that it got to be kind of a joke with the rest of us. Whenever something would go wrong, somebody on the staff would be sure to say, 'There must be a pony in here somewhere.'"

T HE OTHER DAY Josh Gilder, one of my colleagues on the Reagan speechwriting staff, reminded me of Mr. Cho, the barber Josh and I discovered a couple of blocks from the White House. Mr. Cho had come to the United States from somewhere in Southeast Asia—Thailand, as I recall—where it was the custom for a barber to massage each customer's scalp before cutting his hair. When you sat in his chair, Mr. Cho would rub your scalp with the palms of his hands, then knead it with his fingertips. He'd work his way slowly up both sides of your head to your crown, then forward to your eyebrows, then backward to the base of your skull. When Mr. Cho finally finished the massage and began cutting your hair, you'd feel so relaxed that you'd have to grip both arms of the barber chair to keep from sliding onto the floor.

For the six or seven months from the time we discovered him

to the time Mr. Cho moved to a new barbershop in the suburbs, Josh and I found ourselves getting our hair cut almost once a week. Soon we stopped thinking of Mr. Cho as our barber and began thinking of him as our therapist. Our visits to the barbershop amounted to our own modest exercises in stress management. Mr. Cho helped us cope.

On a typical afternoon, for instance, Josh and I might have drafts of two or three speeches spread across our desks. The telephones would be ringing. Members of the National Security Council or the Office of Management and Budget would be pestering us for rewrites. Our boss, Tony Dolan, the chief speechwriter, would be clomping down the marble-tiled hallway in his cowboy boots to ask us whether in writing certain passages we had actually intended to cause pointless trouble with the senior staff or had simply gone out of our minds. When it got to be too much, Josh or I would telephone the other.

"Haircut?"

"Thought you'd never ask."

Josh and I have agreed ever since that only one other event could compare with a visit to Mr. Cho. That was a visit to the Oval Office.

"Reagan's presence was just—I don't know, remarkable," Josh says. "We'd go in there, all worked up over staff wars or the way the researchers weren't doing their work. We might even have been worked up over something important for a change, like the Sandinistas or the situation in the Middle East. Then Reagan would calm us right down. He was just so sweet and serene. A few minutes with the guy were just as good as one of Mr. Cho's massages. Remember?"

Would I ever forget? Ronald Reagan's serenity taught me one of the most important lessons of my life.

17

——— *The Un-Sheen* ———

For a long time, though, I just couldn't figure it out. I made a mistake about Reagan that you'll understand immediately if you've ever watched the television program *The West Wing*. The program does a good job of portraying the intensity in the White House—people who work there really do look serious, speak earnestly, and spend half their days taking urgent telephone calls and the other half hurrying to vital meetings. My mistake lay in assuming that the intensity must reach a peak or climax in the person of the President. If the people who worked for him were driven and harried, it stood to reason that the President himself must be the most driven and harried of all. *The West Wing* makes the same assumption. Just look at the way Martin Sheen plays the role of chief executive. The man's anguished soul searching never lets up.

Yet in the Reagan White House, the intensity didn't peak in the person of the President. It evaporated. Where Sheen often appears rumpled, Reagan always appeared immaculate, his shirt unwrinkled, his tie snugly knotted, a knife-edge crease in his trousers, his shoes gleaming. Where Sheen is often distracted and on edge, Reagan was always utterly relaxed, giving anyone who ever met him the feeling that he had all the time in the world. (People lost track of time in the Oval Office so easily that Jim Kuhn, Reagan's personal aide, frequently had to step inside to bring meetings to an end.) Where Sheen often raises his voice, engaging in at least one shouting match per episode, Reagan spoke gently. A shouting match? Reagan? Unthinkable.

The day after Josh Gilder joined the White House staff—the

18

date, early 1983, makes this incident telling—I walked Josh through the White House complex to show him around. When we reached the end of the hallway that runs through the East Wing, a Secret Service agent stopped us, asking us to step to one side, standard procedure when the President himself was about to appear. A moment later the door at the far end of the hallway opened. The President and an entourage consisting of a couple of Secret Service agents and Edwin Meese III, then counselor to the President, entered, walking in our direction. If the President had worn a grave expression or walked with his shoulders stooped, I wouldn't have been surprised. Although the economy had begun to edge its way out of the recession of 1981 and 1982, one of the worst since World War II, unemployment remained at more than 10 percent. American plans to deploy nuclear missiles in Western Europe—the deployment was intended to offset nuclear missiles the Soviet Union had already deployed in Eastern Europe—had prompted massive demonstrations in the United States and in the United Kingdom, West Germany, and elsewhere in Europe. (Traveling in Europe with the Vice President that summer, I would experience one of these protests at first hand. As the Vice President's motorcade drove through Krefeld, West Germany, spectators began to whistle and jeer, then pelted the motorcade with rocks.) The President's job approval rating, once well over 60 percent, had sunk to under 40 percent.

As the entourage approached, I could hear Meese, briefing the President, refer repeatedly to Laos. I had no idea what Meese was saying about Laos—I'd only been able to pick out that one word—but he wouldn't have been talking about a small country on the other side of the world if the news were good. Yet instead of looking grave or stooped, I could see as he drew closer, the President looked—well, he looked marvelous. Flawlessly groomed and dressed. Rested and relaxed. Alert. At seventy-two, an age when a lot of people hobble from place to place complaining about their arthritis, he even had a spring in his step. Just before the entourage disappeared through the door at our end of the hallway, the President noticed Josh and me. Although he as yet had no reason to know who we were—Josh and I were both still working for the Vice President—he nodded, then grinned. If Reagan felt the weight of the world on his shoulders, the sensation gave him a kick.

—— *Some Luck* ——

How DID Reagan do it? He was supposed to play the role of the lonely and tortured soul. Instead he remained lighthearted. He never cursed or moped, like one of his predecessors, Richard Nixon, or withdrew into isolation for days at a time at Camp David, like another of his predecessors, Jimmy Carter. (For that matter, he never put on twenty pounds, then took it off, then put it back on, or fell into the habit of placing telephone calls to cronies in the small hours of the morning, like one of his successors, Bill Clinton.) Josh Gilder and I were only speechwriters. Yet the pressure got to us so often that in order to cope we had to keep running out to have our heads rubbed. Reagan never had a Thai barber. How did he remain so serene?

In answering this question I made a second mistake about Reagan. I assumed he'd been lucky.

It's not that I was wrong, exactly. Ronald Reagan *had* been lucky. Born in 1911 and raised in a series of obscure towns in northern Illinois, including Dixon, the hamlet in which his family finally settled, by 1932 Reagan had talked his way into a job as a radio sports announcer that made him a celebrity in much of the Midwest. Then in 1937, when he was assigned to follow the Chicago Cubs to their spring training camp on Catalina Island, Reagan had visited Hollywood, where he had looked up a minor actress he had known when she worked in radio in the Midwest. The actress had introduced him to an agent, the agent had arranged for him to take a screen test, and on the basis of that single screen test Warner Brothers had offered him a contract. In the middle of the Great Depression, when as much as a quarter of the American workforce was unemployed, Reagan had lucked into a job that would make him a movie star.

Reagan had stayed lucky all his life. He had gone from a successful career in movies to a successful career in television, hosting *General Electric Theater* and *Death Valley Days*, and then to a successful career in politics, serving two terms as governor of California before capturing the White House. Why was Reagan serene? When you get lucky young, then stay lucky all your life, I thought, you get used to the idea that things are always going to work out for you.

Correct as far as it went, my thinking left a lot out. Thinking? I'm flattering myself. During my first couple of years at the White House I was too busy to think. Since I had no experience as a speechwriter, I spent all my time struggling to learn my job. Any spare energy I had went into pretending that I already knew what I was doing, not into plumbing Ronald Reagan's character. Only when I stopped suspecting that I might be fired at any moment did I begin to give any real consideration to the forces that had shaped Reagan's outlook. I began getting to know Reagan associates, including his longtime political aides, Edwin Meese III and Franklyn "Lyn" Nofziger. I became acquainted with A. C. Lyles, who had been a pal of the President's ever since Reagan's first year under contract to Warner Brothers; During A.C.'s frequent trips from Hollywood, where he remained a producer at Paramount, to Washington, where he'd visit "Ronnie," I'd sometimes join him for lunch in the White House mess. Able to lead an ordinary life after office hours at last—for many months I was so exhausted when I got home from the office that I'd collapse in front of the television—I began reading everything about Reagan that I could find, including his autobiography, *Where's the Rest of Me?* Now that I was indeed plumbing Reagan's character, I quickly recognized my mistake. If you wanted to understand the President, I saw, toting up his good luck wouldn't get you very far. You also had to consider how much bad luck he'd suffered.

Three of Reagan's misfortunes came to strike me as central. The first: His father, Jack Reagan, a shoe salesman, was a drunk.

In *Where's the Rest of Me?* Reagan describes two instances of his father's drunkenness. The first took place when Reagan was eleven. He came home one day to find his father "flat on his back on the front porch."

> He was drunk, dead to the world. I stood over him for a minute or two. I wanted to let myself in the house and go to bed and pretend he wasn't there. Oh, I wasn't ignorant of his weakness. I don't know at what age I knew what the occasional absences or the loud voices in the night meant, but up till now my mother, Nelle, or my brother [Neil, nicknamed "Moon," who was about two and a half years older than Reagan] handled the situation and I was a child in bed with the privilege of pretending to sleep. . . .
>
> I bent over him, smelling the sharp odor of whiskey from the speakeasy. I got a fistful of his overcoat. Opening

22

the door, I managed to drag him inside and get him to bed.

The second took place when Reagan was a young man. His father reeled home one evening, too drunk to remember what he'd done with the family car. Reagan had to search for the vehicle, finally coming across the car in the middle of the road, the engine still running.

The child, dragging his father into the house or listening, huddled under the covers, to voices in the night. The young man, forced to search for the family car, possibly his family's most valuable possession, only to discover that his father had left it where it might easily have been struck by another vehicle or stolen. If I found these incidents disturbing, I kept thinking, what must they have been like for Reagan himself?

When he dragged his father to bed, Reagan writes, it was "the first time" he'd found Jack unconscious. How many times did the scene repeat itself? How much of the family's income, always modest, did Reagan see his father squander? How many of the half-dozen moves his family made during his childhood took place because his father had gotten fired for missing work when he went on a binge?

The small towns of the Midwest in which Reagan was raised would have been dominated by old-stock Protestants, many of whom would have needed only the slightest prompting to look down their noses at a family headed by a first-generation Irish Catholic such as Reagan's father (Reagan's mother, Nelle, was a Protestant of Scots and English ancestry). Jack Reagan gave his neighbors more than the slightest prompting. And he did so during Prohibition. Remaining in effect from 1920, when Ronald

Reagan was nine, until 1933, when he was twenty-two, the Eighteenth Amendment to the Constitution banned the manufacture, transportation, or sale of alcoholic beverages. During his boyhood and college years, Reagan would thus have seen his father not only as an alcoholic but as a petty criminal.

How much shame did Reagan experience? How much humiliation? As I was pondering these questions I gained an insight into them from an unexpected source, my own father. Talking over family history with me one day when I was visiting our home in upstate New York, my father paused. "There's something you might as well know," he said. "My grandfather, your great-grandfather, was an alcoholic."

My father knew little about the man. When he was growing up, he explained, no one spoke about him. My father wasn't even certain when he had died. From conversations he'd overheard, my father had pieced together the story. There had once been some money in the family—not much, probably, but some—but his grandfather had squandered it, then deserted his wife, leaving her to support three children, one of them my father's own father, on her own. As best he could recall, my father had been permitted to meet his grandfather only once—he still had a dim memory, dating from his earliest boyhood, of lying on a sled as his grandfather pulled him across the snow. "I remember liking him," my father said. "He seemed kind. But that's all that I can tell you."

By the time I was born, my father's grandfather was probably dead. Why had my father waited until I was in my twenties to tell me about him? Because as a boy he had recognized that adults saw his grandfather as a disgrace, and now, all these years later, he himself still felt ashamed. "I didn't want to burden you with it," he explained.

I stood at a remove of three generations from the alcoholic in our family. But I still felt an aftereffect, a faint echo, of shame. As the very son of an alcoholic, I saw, Reagan must have suffered a particular kind of hell.

Little as he says about his father's alcoholism in *Where's the Rest of Me?*, Reagan devotes even less attention to the second misfortune, his divorce from his first wife, Jane Wyman. "I've never discussed what happened," he writes, dispensing with the matter in a sentence, "and I have no intention of doing so now." Now in her late eighties, Wyman, for her part, has always refused to comment on her relationship with Reagan. (Look over the websites devoted to Wyman and you're almost as likely to see tributes to her dignified silence as to her acting.)

Reagan and Wyman met on the set of the 1938 release *Brother Rat*. A beautiful, spirited actress, Wyman was about to marry a man named, improbably enough, Myron Futterman. Although she went ahead with the wedding, she soon filed for divorce, abruptly bringing the marriage, her second, to an end. In 1940, Reagan, then almost 29, and Wyman, 23, married. Their daughter, Maureen, was born in 1941. Four years later, in 1945, they adopted Michael, then nearly one. A 1946 photo shows Reagan, Wyman, Maureen, and Michael holding hands, striding toward the camera. Reagan is beaming. It is easy enough to imagine why. He was an established actor, a homeowner, and a husband and father who was providing for his family on a scale even the wealthiest citizens back in Dixon would have been unable to match.

In 1948, Wyman divorced him.

Why? Wyman claimed mental cruelty, but that was merely legal boilerplate. The likeliest explanation may be that Wyman decided her husband was holding her back. If so, the marriage amounted to a casualty of World War II. Reagan's career was still on the rise when the war broke out. Yet in serving from 1942 to

1945 in the Army Air Force—his poor eyesight made him ineligible for combat duty, and he was assigned to the First Motion Picture Unit, a publicity unit, in Culver City, California, where he helped to produce more than 400 training and morale films— Reagan put his career on hold. During the same period Wyman appeared in one picture after another, rising from a minor role in the 1942 light comedy *My Favorite Spy* to a major role in the 1945 classic *The Lost Weekend.* After the war, Reagan picked up his career where it had left off—but only where it had left off, failing to break through to stardom of the first rank. Wyman, by contrast, flourished, landing a series of big roles, including the part of a deaf-mute rape victim in the 1948 release *Johnny Belinda*, for which she received an Academy Award. Why remain married to a second-rate actor, Wyman may have reasoned, now that she was one of the most important leading ladies in town?

Reagan was devastated. "That was a very hard time for Ronnie," A. C. Lyles says. "Very hard." Reagan wrote to the Hollywood columnist Louella Parsons that Wyman was "sick and nervous and not herself," as if the only version of reality Reagan could accept was that Wyman had sought the divorce because she was ill. Reagan found his loneliness difficult to accept. "All of us, I suppose," he writes in *Where's the Rest of Me?*, "have a lonely inner world of our own, but I didn't want to admit to mine. . . . Looking back, because at the time I wouldn't admit it to myself, I wanted to care for someone."

Even though most of the Reagan hands I got to know at the White House—Ed Meese, Lyn Nofziger, Bill Clark, and others— became acquainted with the President after he entered politics in the mid-1960s, almost two decades after the end of his first marriage, all agreed that the divorce represented the worst trauma of

Reagan's life. Conducting interviews for this book, I spoke to one longtime Reagan associate who refused to discuss the matter until I'd turned off my tape recorder, promising to keep his remarks off the record. Reagan had often told him, he said, that the end of his first marriage represented the worst personal failing of his life. At first I couldn't figure it out. Go off the record to say nothing more remarkable than that? Then I realized that my interview subject was merely demonstrating respect for his old friend. If Reagan had always found his divorce too painful to mention in public, then my interview subject wasn't going to put anything about it on the record, either.

The third misfortune: About the time he remarried, Reagan found his acting career coming to an end.

Even after I began reading up on Reagan, I assumed his transition from movies to television had been easy. In 1953, after all, he had completed work on one of his last motion pictures, *Prisoner of War*, and then in 1954 he had begun hosting the television program *General Electric Theater*, one of the most popular on the air. Then I had lunch with Charlton Heston. (I'd attended a Kennedy Center performance of *The Caine Mutiny* in which Heston was appearing—one of the perks of working at the White House was free seating in the presidential box—and realized that Heston might be able to give me some insight into Reagan's years in Hollywood. On a long shot I sent the actor an invitation to join me in the White House mess. I figured the man who had played Ben-Hur would probably refuse to associate with a lowly speech-writer, but I also figured I had nothing to lose. To my surprise, Ben-Hur said yes.) Over lunch Heston told me a great deal about Reagan's tenure as President of the Screen Actors Guild— although always charming, Heston said, Reagan had proven such a skillful negotiator that he had brought the studios "to their knees"—but what struck me were Heston's comments about television. During the 1950s, he explained, Hollywood had looked down on the medium.

"When I went into motion pictures," Heston said when I spoke to him once again last year,* "the studios told us, 'You absolutely cannot do television in any form anywhere. It's just a gimmick. It's going to pass away.' Those may have been some of the stupidest remarks in history. But that's the way Hollywood thought in those days."

A gimmick? A fad that would pass away? If that was the way Hollywood thought, I wondered after lunch, then why had Reagan gone into television in the first place? Reexamining *Where's the Rest of Me?*, I found that Reagan's transition from movies to television hadn't been easy after all. Describing the period just before he began hosting *General Electric Theater*, for example, Reagan notes that the studios had begun sending him second-rate material. "For fourteen months I turned down such scripts as came my way." Fourteen months without a role? I hadn't thought of it this way before, but what must that have meant to a man who had appeared in seventeen pictures during his first three years in Hollywood? At one point, Reagan writes, he accepted a two-week gig as the emcee of a floor show in Las Vegas. Reagan passes this experience off as a lark, and the first time I'd read the passage I'd accepted his description at face value. Now I had my doubts. A former President of the Screen Actors

---

* Like Ronald Reagan, Charlton Heston came from the Midwest, served six terms as President of the Screen Actors Guild, and proved to be that Hollywood rarity, an outspoken conservative. A couple of months after I spoke with him, Heston made public a final parallel between himself and Reagan, announcing that he had Alzheimer's disease.

Guild introducing acts in Las Vegas? Reagan, I concluded, had gone into television for one reason. He'd needed the work.

This period in Reagan's life, his wife has since revealed, was even more difficult than I thought. In *I Love You, Ronnie*, the book of love letters that she published in 2000, Mrs. Reagan writes that she and her husband "couldn't afford to furnish our living room." Although Mrs. Reagan gave up acting when she married—"I wanted," she writes, "to be a wife and mother"—she soon returned to work. "I took a part in a film called *Donovan's Brain*, a science-fiction picture in which Lew Ayres plays a scientist who tries to keep a brain alive and is taken over by it." Recently remarried, and the father, with the birth of Patricia in 1952, of an infant daughter, Ronald Reagan produced so little income that his wife had to supplement his earnings by taking a role in a sci-fi flick.

His father was a drunk, his first wife divorced him, and when he was still in his early forties he found the acting career that had given him fame and wealth coming to an end. Yet when I encountered Ronald Reagan, as he was serenely receiving speechwriters in the Oval Office or cheerfully greeting them in White House hallways, he was the sweetest and most untroubled man I'd ever met. It took me a while to reconcile the contradiction between the man Reagan had been—the man, that is, who had suffered his full share of misfortunes—with the chief executive so utterly relaxed and at peace that far from conveying any sense of the burdens of his office, he always made your own burdens feel lighter. When I did, I saw that the contradiction wasn't a contradiction at all. It was a paradox, a contradiction that resolved itself once you looked at it the right way. The right way to look at Ronald Reagan? As proof that we possess free will.

—— *Deep Structure* ——

WHEN I LOOK back on myself during my years at the White House, I see that I was hungry for knowledge of every kind. Why? I can only tell you that I'd grown up in a town in upstate New York that was far removed from any major city; that I'd spent half a dozen years in college and graduate school, narrow environments in their own way; and that when I emerged, blinking with astonishment, into the Washington, D.C. of the 1980s, I felt like a medieval peasant or monk who suddenly found himself in Rome. The big city seemed so dreamlike and wonderful that I wanted to clutch it to my chest. If I understood it better—Reagan, politics, the art of writing, for that matter the ideals of Western civilization itself—then perhaps, I thought, I could make the experience last, finding a place for myself in this amazing new world before anyone found out I'd arrived on a fluke. When I mentioned to a friend over a beer one night that I was interested in theology—in an administration opposed to the "evil empire," I figured, I ought to read up on good and evil—he introduced me to the theological adviser to the archbishop of Washington, Father Lorenzo Albacete, a big, unkempt priest from Puerto Rico. The holder of a doctorate in theology and a master's degree in applied physics, Father Albacete had a deep, melodious voice, an engaging manner, and a proclivity for belly laughs. Two or three times a month I'd drive from the White House into the Maryland suburbs, pick up Father Albacete at his rectory, then take him to dinner at a fast food restaurant, usually Kentucky Fried Chicken, our favorite. We'd remain until the restaurant closed. As we ate, Father Albacete would talk while I scrawled notes on napkins.

"A lot of people have a mistaken conception of free will," Father Albacete said one night, picking through our bucket of chicken. "They think exercising free will means choosing their own reality. Try hard enough, and you can make yourself rich or famous or beautiful—that kind of thing. Well, man, I'm sorry. But it just ain't so.

"Nobody," he said, pulling out a wing, "gets to choose his own reality. Oh, sure, you can have a little effect on reality here and there. If you diet, you'll weigh less than you will if you eat the way we're eating right now. If you work hard, you'll make more money than if you're lazy. Every so often somebody comes along who even seems to push history itself in one direction or another. But nobody gets to choose his parents. Nobody gets to choose whether he's good looking or ugly or whether he's intelligent or stupid. We all have to take reality as it comes to us—Presidents, popes, all of us."

Ronald Reagan, Father Albacete said, had never chosen to have an alcoholic father, to have his first wife end their marriage, or to have Hollywood's taste in actors shift so markedly in the early 1950s that the studios suddenly wanted fewer leading men like Reagan, who played the good-natured fellow next door, and more leading men like James Dean and Marlon Brando, who played brooding misfits. All that had just happened. That had been reality as it had come to him.

"If the question is not what reality you choose," Father Albacete continued, waving the wing, "then what *is* the question? The question, my friend, is what you choose to *do* with reality. Reagan never permitted his misfortunes to interfere with his development as a human person. Instead he used them. Do you see? All his life Reagan exercised his free will by choosing to seek the good in reality as it came to him."

Reagan might, for example, have grown to resent his alcoholic father. Instead he adopted the attitude of his mother, who believed, as he put it in *Where's the Rest of Me?*, that "we should love and help our father and never condemn him for something that was beyond his control." Reagan might have brooded over his divorce, permitting bitterness to immobilize him. Instead he immersed himself in his work, remaining physically fit while performing his duties as an actor conscientiously. When he had a moment, he devoted it not to moping but to fan mail. "He answered a lot of the fan mail himself," A. C. Lyles says, "usually in longhand." And when he found an opportunity for a happy married life in the person of Nancy Davis, he took it. Reagan might have responded to the end of his acting career by growing dejected and withdrawn. Instead he did what he could to provide for his family, accepting parts in low-budget pictures, appearing in guest spots on television, and even, as we have seen, performing in Las Vegas. When in 1954 General Electric offered him a contract, the terms called for Reagan to serve as a corporate flack, spending days at a time away from home as he toured General Electric plants across the country, and to appear on television, still a disparaged medium. Although the contract with General Electric would transform Reagan's life, providing him with intensive training as a public speaker while enabling him, as the host of *General Electric Theater*, to master television more completely than any other politician of the twentieth century, Reagan could have known none of that at the time. Yet he signed the contract without complaint.

"Look at him," Father Albacete said, finishing the wing. "He may have been the son of a drunk, but in all these experiences

Reagan somehow found what he needed—the strength, the re-
siliency—to become the leader of the most powerful nation on
earth. Bringing good from bad. Why should that be possible?
Because of the deep structure of creation. Because of the way God
himself ordered the universe. Remember Genesis? 'And God saw
that it was good.' What you have in Ronald Reagan is a man who
has made contact of the most direct and intimate kind with divine Providence itself. *That's* why Reagan is so serene."

As Father Albacete completed his meal, I told him Reagan's
favorite story, the pony joke. At the punch line, Father Albacete
tossed his wing onto his plate in amazement. "That's *it*," he said.
"That's *it*. That's the entire anthropology of human existence. You
become a complete person by digging for the pony in the midst
of all the crap life throws at you."

He laughed. "Ronald Reagan, teaching *me* theology. My God."

—— *On the Congo* ——

As I DROVE home that evening, I found myself recalling a
conversation I'd had with Frederick Ahearn, a White House
advance man, when Rick and I had whiled away an afternoon together in the unlikely location of Kinshasa, Zaire (now the
Democratic Republic of the Congo), a couple of years before. Still
on the staff of Vice President Bush, I'd spent the previous ten days
traveling with the Vice President to half a dozen countries in
Africa; Rick, in charge of arrangements for the Vice President's
two days in Zaire, the final stop on Bush's itinerary, had been in
Kinshasa for several days. Although the Vice President had departed that morning, Rick and I wouldn't leave until late that
evening. So we'd gone to the American embassy compound,

climbed into our swimsuits, pulled up a couple of chairs near the pool, and sat in the sun drinking soda as we watched the Congo River slide past.

As we talked, Rick told me about March 30, 1981, the day Reagan was shot. Rick had been with the President—you can see him in all the photographs of the confused scene outside the Washington Hilton. At first, Rick said, everything seemed routine. The President completed a speech on economics before a union group. Then he and his party left the hotel ballroom, made their way upstairs, and departed from the building through a side entrance. Just before climbing into his limousine, the President turned to wave to the small crowd that had gathered to see him. Then the gunman fired.

"I heard *bang bang*," Rick said. "Then there was a pause. Then I heard *bang bang bang bang*. All six rounds were expended in less than two seconds." For an instant Rick became disoriented, unable to understand what was happening. "All the assailant had was a .22-caliber handgun," Rick explained, "but I was familiar with heavier caliber weapons. At first I thought somebody had set off firecrackers in the parking lot as a prank." When Rick recognized that the President was under attack, he lunged forward to help the Secret Service agents shove Reagan into the limousine. "That was when the fourth round hit the door window," Rick said. "The bullet sent up a spray of pulverized glass that hit me in the cheek and sent me backward. I'd always been told that when you're shot in the face it stings like a thousand needles, and now all of a sudden I had a thousand needles in my cheek. I brought my hand up and saw there was no blood. By the time I looked again, they had the President in the limousine and the motorcade was leaving." About this time, Rick later learned, the assailant, a young man named John Hinckley who had opened fire at a distance of only ten feet, was tackled and taken into custody.

Unaware that the President had been wounded, Rick turned his attention to the three men lying on the sidewalk: Timothy McCarthy, a Secret Service agent who had been shot in the chest; Thomas Delahanty, a Washington, D.C., policeman who had been shot in the neck; and James Brady, the White House press secretary, who had been shot in the head. "I knew there were still a lot of Secret Service agents on the scene who would take care of Tim McCarthy, and I figured the D.C. cops would take care of the policeman. But there was nobody to take care of Jim, and I could see right away that he had a head wound. The bullet had hit him in the left side of the forehead." Rick knelt by Brady, talking to the press secretary while using a handkerchief to stanch the bleeding from his head. When an ambulance arrived, Rick climbed in, insisting on remaining with Brady. "The medics wanted to take Jim to the MedSTAR Unit at the Washington Hospital Center," Rick said. "That was way the hell across town, and I'd had enough training to know that speed of treatment is essential with a head wound. The brain starts to swell, and the only place it can expand is downward. That puts pressure on the brainstem, which controls respiration and heart function. So as the brain swells, the patient starts to die. There was no way in the world that this ambulance was going to get to MedSTAR in time for anyone to save Jim Brady. I said, 'Bullshit'—pardon my French—'this man won't make it to MedSTAR. We're going to G.W. Hospital [George Washington University Hospital, only a couple of miles away]. And if you don't like it, I'll open this door and throw your ass right out of the ambulance.' So we hauled down Connecticut Avenue."

Doctors began working on the press secretary the moment the

ambulance reached the hospital. (Brady, like McCarthy and Delahanty, would survive.) Still unaware that the President had been wounded, Rick noticed to his surprise that he was not the only member of the White House staff in the emergency room. "Dave Fischer [then the President's personal aide] and Mike Deaver were standing there inside the emergency room. Each was on a telephone. Deaver was trying to reach the first lady and Fischer was trying to get hold of the Vice President, who was in Texas. I said, 'What are you guys doing here?' Dave said, 'Goddamn it, the President's here.' I said, 'Is he okay?' Dave said, 'We don't know what's wrong with him. He may have had a heart attack. We just don't know.' Until the doctors came out and told us the President was gravely wounded, nobody knew what was going on."

Rick remained at the George Washington University Hospital almost nonstop during all twelve days the President himself remained there. During that first day, Rick helped a Secret Service agent establish a security perimeter, set up an area outside the hospital for press briefings, made the first public announcement that the President had been injured, and met Mrs. Reagan when she arrived, taking her to see her husband. Over the eleven days that followed, Rick coordinated hospital visits by members of the Reagan family, by virtually every member of the White House senior staff, and by half the Cabinet. "As soon as he was well enough, the President started doing business right in his hospital room," Rick said. During his hospital duty, Rick had plenty of time to compare notes with other eyewitnesses, learning just how badly the President had been wounded.

"When the agents pushed him into the limousine," Rick explained, "the President reached out his arms to break his fall. That was when the fifth round was fired." The bullet had struck the left rear panel of the limousine, flattening to about the size of a dime. Then it had traveled along the limousine to slip through the space between the body of the limousine and the open limousine door, entering the President's body through his left armpit. "Jerry Parr, the leader of the [Secret Service] detail, jumped into the limousine on top of the President," Rick said. "At this point the President was lying across the driveshaft in the back of the limousine, and he said, 'Jerry, get off me, I think you've broken my rib.' Jerry helped the President sit up. Then the President coughed. Jerry saw bloody, bright pink foam on the President's lips—the color showed that the blood was fully oxygenated, indicating a lung wound—so he told the driver to divert the limousine to G. W. Hospital. The President got out of the limousine on his own, but when he was halfway through the hospital door his legs gave way."

Although the medical team had recognized at once that Reagan was losing blood, it had had trouble locating the wound. "It wasn't until they cut his clothes off that they found that little puncture mark under his armpit, and by then his lung was collapsing," Rick said. During surgery, the team had discovered the bullet just an inch from the President's heart. "They figured out afterward that it struck one of his ribs, then tunneled through his lung," Rick said. "Just an inch from his heart. Can you imagine that?" Less than three months old, the administration of Ronald Reagan had nearly come to an end. "It was close," Rick said as we watched the sun play on the Congo River. "Real close."

T HE ATTEMPT ON Reagan's life, I saw, represented a kind of test case—perhaps the ultimate test case. If Reagan had been serene because he'd been lucky, then the bullet that pierced his lung, lodging near his heart, would have shattered his equanimity, demonstrating that his long run of luck had come to an end. He'd have become a shrunken figure, a lesser man. Of course that hadn't happened. Instead he'd become a larger figure, a still more commanding man. Close to death, he had made jokes, telling his surgeons he hoped they were "all Republicans" and famously explaining to his wife, "Honey, I forgot to duck." ("He really did tell Mrs. Reagan he forgot to duck," Rick Ahearn had told me in Zaire. "I was there when it happened.") Discharged from the hospital, he had returned to the White House to pursue his agenda, permitting himself less than three weeks before traveling to the Capitol to address a joint session of Congress.

Reagan wasn't serene because he'd been lucky. He was serene because he'd been *un*lucky, learning to bring good from bad so thoroughly that he retained his equanimity even after an attempt on his life.

JOURNAL ENTRY, NOVEMBER 2002:

After the shooting, Reagan often told people that whatever time he had left belonged to "the man upstairs." I never doubted he meant it, but I did sometimes wonder how much *he meant it. Was he merely expressing a vague sense of piety, or something deeper?*

Today Lydia {my assistant} pointed out an entry in the President's diary that she'd come across in one of the library books she was using as a reference. "I know it's going to be a long recovery," Reagan wrote on his first night in the White House after leaving the hospital. "Whatever happens now I owe my life to God and will try to serve Him every way I can."

It was just as Father Albacete said over dinner at Kentucky Fried Chicken. Ronald Reagan had made contact of the most direct and intimate kind with divine Providence itself.

—— *Pony to the Rescue* ——

I'VE NEVER endured anything like Ronald Reagan's misfortunes. And although I've tried, I've never achieved anything like his serenity, either. But I can't tell you how often I've thought about the pony in the dung heap, even so. The story combines two truths. We possess free will—if we choose to do so, we can dig. And even the most painful situation contains a great deal of good. That's not wishful thinking. It reflects, as Father Albacete put it, the deep structure of creation. Let me describe a couple of instances in which I've put the lesson of the pony to use.

After leaving the White House, I attended business school at Stanford; then, as I was about to graduate, I accepted a job with the News Corporation, the company founded by Rupert Murdoch. Murdoch, who hired me himself, wanted me to help him build a cable news operation to compete with CNN. Eleven months after I went to work, one of Murdoch's assistants called me to his office. The company, he explained, had decided to delay the new television venture—the recession of the early 1990s had denied the News Corporation the cash the venture would have required*—so I'd need to find a new job.

I'd been laid off.

At first I just didn't understand. I'd thought that layoffs only happened to blue-collar workers. My collar was as white as they got. I'd had a fancy education. I'd worked at the White House. Wasn't there some *law* against laying off people like me?

I spent the first few weeks feeling low. My wife had just given

---

*The venture, Fox News, was indeed eventually launched—but not for another six years.

birth to our first child, but instead of living in some comfort in our own place—we'd rented an apartment in Manhattan—the three of us found ourselves moving in with my in-laws. I sent out a few dozen résumés but otherwise sat around feeling sorry for myself.

Then I remembered the pony in the dung heap. If Reagan had been in my position, I recognized, he'd have found a way to bring some good out of it. Determined to follow his example, I set up an old card table in my in-laws' basement, then started work on a project that had been in the back of my mind ever since I'd first considered going to business school: a book about business school itself.

During my final year at the White House, you see, I'd read *One L*, a book about Harvard Law School that had helped me to see law school wasn't for me. But when I'd looked for a similar book about business school, I'd discovered that none existed. If I did attend business school, I'd thought, I ought to take a lot of notes, then compose an account myself. Working for Rupert Murdoch, I'd had too little time to write a book. Now I had nothing but time.

I began the book with my last day at the White House ("You be careful out there," Reagan had said with a grin. "That Stanford faculty is a little left-leaning"), cut to my first week of classes, when I found that I was unable to keep up, even in a course called "Microeconomics for Poets," and took it from there. *Snapshots From Hell: The Making of an MBA* was published in 1994.

The second instance is recurring—a permanent circumstance. Like a lot of people who write for a living, I find that a bad day at the word processor can put me in a funk. (My wife has been known to tell me to stay at the office until I cheer up because she has enough children at home already.) After producing page after page of work that appears worthless, I'll find myself unable to look at it. Instead I'll surf the web or play solitaire, wasting time.

Then I'll recall the lesson of the pony—I really will—and force myself to reread the day's output. Sometimes, I suppose I'd better admit, I find myself throwing the material in the wastebasket. But quite often I'll find a few sentences good enough to keep. Plucking them out, I'll begin rewriting. Before long, to my surprise—for some reason, this never ceases to astonish me—I'll have a page or two that are decent.

Yes, I know. Whereas in the first instance I'd found myself laid off, this second instance involves only workaday troubles. But that's my point. Exercising my own free will, choosing to search for the good in bad situations—I've called the lesson of the pony in the dung heap to mind so often that it has become part of my habitual outlook, helping me face not only crises but the petty difficulties of everyday life. I'm still not particularly optimistic. If I see a glass of water, I'll think it's half empty, not half full. Then I'll start wondering how long it's been standing there and what kind of bacteria it might contain. But the lesson I learned from Ronald Reagan has made me a much more serene person myself.

Lately I've found myself teaching the lesson to my children. The disappointments they're always suffering—losing a Little League game, getting a bad grade on a test—can put them into funks of their own. "When life buries you," I'll say, "start digging. You'll learn a lot about your own strength and resilience, and about the goodness of creation itself." It's amazing how willing to accept these sermons the children always prove. And I? By climbing into their little heaps with them, I'm brought into their lives, becoming a better father. Even a child-sized mess can hold a pony.

Two

THE POSTHOLE DIGGER

Do Your Work

My assignments for the next couple of days:

Remarks for a women's event being billed as a "Birthday Party for Susan B. Anthony," a two-minute taping for the Automotive Service Industry Association, and a toast for a dinner benefiting the Princess Grace Foundation. I suppose if I approached these assignments in the right frame of mind I'd see they represented some sort of service to the republic. But Susan B. Anthony? A dour suffragette who's been dead for three quarters of a century? How am I supposed to get twenty minutes out of that? The Automotive Service Industry Association? How can I write anything substantive, moving, entertaining, sprightly, or witty . . . about spark plugs and windshield wipers? The Princess Grace Foundation? Grace Kelly was a pretty good actress—I liked her a lot in Rear

Window—*but* Princess *Grace? I mean, Monaco comprises fewer acres than Central Park, and I'm supposed to write about it as if it* mattered?

A nineteenth-century activist, auto supplies, and a princess from the tiniest country on earth. And I thought that when I joined this staff I'd be writing about the great issues of the day.

——*Toil and Trouble* ——

WORKING AT the White House sometimes produced a kind of a wild high—I've never since experienced quite the thrill I used to get from turning on the evening news to watch a clip of the President delivering a speech I'd written—but most of the time, working at the White House produced only a sense of the work itself. A job, even with the President, was a job. At least that's the way it felt to us speechwriters.

Our lives, I found, broke down into thirds. On each speech we'd spend about a third of our time on research, plowing through background reading that included materials on the audience the President would be addressing and on relevant aspects of administration policy. Clark Judge performed so much background reading that the researchers used to complain about the sheer weight of the materials he was always asking them to fetch. Tony Dolan performed a lot of background reading himself, always searching for material that would enable him to infuse his speeches with a sense of history. Poking around in the Reagan Library not long ago, I found the research files that Tony put together when he drafted the President's 1982 address to the British Parliament. The files, which ran to several hundred pages, included articles entitled "Classical Civilization," "The Decline of the City-State," "Spanish Empire," and "Summary of Louis XIV's Reign."

We'd spend another third of our time writing. Writing, I quickly learned, was writing. Reminding yourself that you occupied an office with a sixteen-foot ceiling and a marble fireplace—our offices were in the Old Executive Office Building (now called the Eisenhower Executive Office Building), the extravagant nineteenth-century building across West Executive Avenue from the White House itself—didn't do much good when you found yourself staring at the blank screen of a word processor as a deadline approached. In the throes of composition, Josh Gilder would wander the halls of the OEOB, going nowhere, speaking to no one, looking stricken. Dana Rohrabacher and I would find excuses for dropping by the offices of each of the other speechwriters, describing our assignments to our colleagues in the vain hope that somebody would be able to suggest a few good paragraphs off the top of his head. Tony Dolan would simply hunker down at his word processor, chewing cigars, for hours at a time, often working into the small hours of the morning. If anyone proved unwise enough to disturb him, Tony would let out a bark like a bull seal.

We'd devote the final third of our time to staffing, the process in which each speech was circulated to the senior staff and certain Cabinet agencies. Almost always, disputes would arise. Officials marking up a foreign policy speech at the State Department and Pentagon, for instance, might insert contradictory comments, forcing the speechwriter to spend a lot of time on the telephone, persuading the officials to sort out their differences. Even when staffing went smoothly, at the end of the process you'd find your desk buried beneath a dozen copies of a speech, each of which had been marked up by a different person. You'd have to make your way laboriously through each marked-up copy to assemble a final draft.

On big speeches, speeches you knew would prove important, a sense of excitement would carry you along through all three steps. But for every major address the President delivered from the Oval Office, he delivered dozens of routine remarks in the Rose Garden, the East Room, and Room 450, the auditorium on the fourth floor of the OEOB. In my first year on the President's staff I drafted remarks for the American Legion Girls Nation; for the American Legion Boys Nation; for a ceremony to congratulate Doug Flutie, the winner of the Heisman Trophy; for a Boy Scouts award ceremony; and for a parade at Epcot Center in Disney World. It wasn't excitement that carried me along. It was the knowledge that I needed a couple of paychecks each month to cover my rent.

Once I joined the President's staff I began receiving calls from headhunters whenever a corporate speechwriting job opened up, and soon I had interviewed for positions at Pepsi, North American Rockwell, Holiday Inn, Bain and Company, and General Electric. I turned them all down. I'm not saying I found the work at the White House so onerous I wanted to leave, in other words, only that I found the work—well, work.

Which brings me to my point. When I landed in the White House, I didn't understand work. I knew how to perform it, of course—if I hadn't, I'd have been fired as quickly as I'd been hired. But I had no idea what to make of it.

Where did work fit in the well-lived life? In the home in which I'd been raised, that had never been a question. My mother had grown up on a farm where she had watched both her parents rise before dawn, her father to milk the cows, her mother to split wood, fire up the big cast-iron stove, and make breakfast. My father had been raised in a house located within walking distance of the warehouse in which his father worked and the shoe factory that employed his mother. As adults, my parents had worked hard all their lives, my father in a series of unglamorous jobs because he had a family to support, my mother supplementing our family income as a portrait painter. In our home, work wasn't something you thought about. Work was something you just did.

In college I'd encountered a different attitude, for the first time meeting people who would never have to work. By an accident of the alphabet, for instance, the mailbox next to mine belonged to Peter Rockefeller, a grandson of Nelson. Once I asked Peter about his career plans. He replied without a trace of irony that he intended to become "a captain of industry." I found myself turning that answer over in my mind for days. To people such as Peter Rockefeller, I saw, work was a kind of adornment. You decided to become a captain of industry instead of a financier or a politician in the same way you might have your architect design a mansion in the style of a French château instead of an English country house or a Venetian palazzo.

At Oxford I'd encountered a still different attitude, falling in with a group of young aristocrats—my friends had included a count, a viscount, the younger son of a baron, and the nephew of a duke. Whereas Peter Rockefeller at least envisioned a place in his life for work, my fellow Oxonians would have preferred to avoid work altogether. To them, the second-best life involved making a fortune, then retiring at forty to play polo. The best life? Inheriting a fortune in the first place.

I wasn't in the position of either the Peter Rockefellers or the young aristocrats of this world, needless to say. With neither wealth nor connections, I had to make my own way. Yet as I endured the drudgery of composing remarks for the American Legion Girls Nation or a speech for Disney World, I'd find myself wondering, Shouldn't I be thinking bigger? Not that I actually had any big ideas, you understand. Turn myself into a captain of industry? How? Start wearing wingtips instead of loafers? Yet I couldn't escape the thought that the way I looked at work was somehow inadequate. And then there was the question of money. Shouldn't I be trying to make a lot more of it? I had nothing to support but my pizza habit, I'll grant you. But while I was collecting government paychecks, my Oxford friends were all in the London financial district, getting rich.

You can see my trouble. I had a job that I liked well enough—and that, whenever I was working on a big speech, I just loved. But I kept supposing I ought to feel discontented. What, I kept trying to figure out, was the right way to look at work?

Ronald Reagan answered my question.

To show you how, I'll need to give you an overview of his economic policies. Bear with me here.

—— *Before and After* ——

All you really need to know to grasp the extent of Ronald Reagan's economic achievements appears in the two sets of charts below. Think of them as "before" and "after" pictures.

The first two charts, the "before" pictures, illustrate the performance of the American economy in the decade before Ronald Reagan became President. The lines that should be going up are

BEFORE

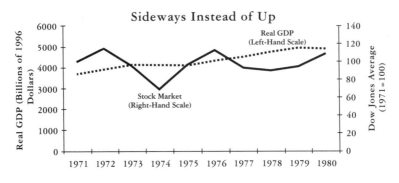

Sideways Instead of Up

Up Instead of Down

going sideways, while the lines that should be going down are going up. The growth rate during the decade was anemic. The Dow Jones Industrial Average merely sputtered along—if you'd invested a dollar in a fund tied to the Dow in 1971, then in 1980 your dollar would have grown to only a dollar and change. By the end of the decade, the number of unemployed Americans had increased by more than two and a half million, inflation had climbed to the highest levels in more than twenty years, and interest rates had soared to levels that were among the highest of the twentieth century. (By January 1981, indeed, the prime rate had soared to the highest level since the Civil War.)

The second set of charts, the "after" pictures, illustrates the performance of the economy during the eight years after Ronald Reagan became President. At first, you'll notice, certain things

AFTER

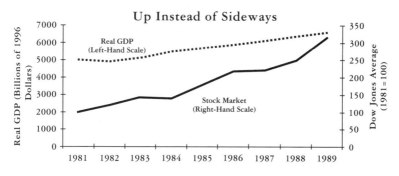

Up Instead of Sideways

Down Instead of Up

got worse instead of better—soon after Reagan took office the economy recorded its worst decline since the Great Depression, sending real GDP down and unemployment up. But then the recession ended and Reagan's economic policies—tax cuts, a reduction in the growth of federal regulations, and support for the Federal Reserve in its efforts to check inflation—began to take effect.

From then on, all the lines that should be going up are going up, while all the lines that should be going down are going down. The economy boomed. From 1981 to 1989 the Dow Jones Industrial Average climbed from under 900 to over 2700—if you'd invested a dollar in a fund tied to the Dow in 1981, by the time Reagan left office in 1989 your dollar would have grown to more than three dollars. By 1989 the number of unemployed Americans was more than 1.7 million lower than it had been when Reagan entered the White House, while inflation and interest rates had plummeted.

Although the charts end in 1989, the year Reagan returned to private life, the expansion that he began continued. As of this writing we can look back on twenty-one years of economic growth marred by only seven quarters of contraction (three in 1990 and 1991, during the crisis that followed the Iraqi invasion of Kuwait, one in 1993, and three more in 2001, after the bursting of the dot-com bubble). Reagan not only turned the economy around so dramatically that you can instantly grasp the difference in "before" and "after" pictures, he established the conditions that permitted economic growth to continue for many years thereafter. No other President in the second half of the twentieth century can claim an economic achievement that equals it. That is my opinion, of course. But I'm not the only one who holds it. While working on this book I thought to ask two Nobel Prize–winning economists, Milton Friedman and Gary Becker, which President in the last fifty years had done the most for the economy. Both answered without hesitation, "Reagan."

How did Reagan do it?

——*Foiling the Gag*——

B Y BEING a stubborn cuss.

Over the life of any administration, economic policy amounts to the sum of hundreds of meetings, speeches, and decisions. But watch Reagan in just three scenes.

The Posthole Digger

*Here is a dollar such as you earned, spent, or saved in 1960.
And here is a quarter, a dime, and a penny—36 cents. That's
what this 1960 dollar is worth today.*
            —Ronald Reagan, address to the nation on the economy,
                                            February 5, 1981

From time to time, the President and the chairman of the
Federal Reserve, Paul Volcker, lunched together. For the first of
our three scenes, picture one of these lunches as it might have
taken place in 1981 or 1982. Over soup and crackers, and, for
dessert, Jell-O—Reagan ate lightly at midday—Volcker briefs
the President. First Volcker reviews the Fed's dramatic reduction
in the growth of the money supply, a policy Volcker instituted to
combat inflation. Then Volcker discusses the deepening recession
that has emerged as a result. Reagan shakes his head in dismay as
he listens. Concluding the lunch, the President says simply,
"Well, Paul, use your judgment and keep at it."

An anodyne scene, yet no less an authority than Milton
Friedman considers it historic. Why? Because of what the
President *didn't* say. "No other President," Friedman argues,
"would have stood by and let Volcker push the economy into recession by restricting the money supply so sharply."

As the recession developed, it imposed a heavy political cost
on Reagan. His job approval rating slid, and in the midterm elec-
tions of 1982 Reagan's party, the GOP, lost twenty-six seats in the
House of Representatives. Since Volcker could have alleviated
the recession merely by easing up on the money supply, many in
the White House wanted the President to pressure Volcker to do
just that. As Ed Rollins, the President's chief political adviser,
told me, "There's only one man who can cost the President re-

election all by himself, and that's the chairman of the Federal Reserve. We need to control that guy."

A President pressure a Fed chairman? Today the idea sounds ridiculous. Reagan's successors, Presidents George H. W. Bush and Bill Clinton, never pressured the Fed chairman, who is by statute independent of the White House. Why should it have been possible for Reagan to do so? Because when Reagan took office there was a long history of Fed chairmen proving only too willing to do whatever the White House bade. As Richard Nixon was running for reelection, for example, Fed chairman Arthur Burns produced a massive expansion of the money supply in a brazen ploy to help Nixon. (When the Fed expands the money supply too rapidly, the effect over the long term is always inflation. But in the short term an expansion can create at least the illusion of economic growth, boosting a President's popularity.) Volcker himself had proven capable of putting the Fed at the service of partisan politics. After pursuing a restrictive monetary policy early in 1980, Vocker had reversed himself in the spring, engaging in one of the most rapid expansions of the money supply since World War II—an expansion that would stimulate the economy in time for the presidential election that autumn. "Volcker was playing games," says Milton Friedman, whose classic *Monetary History of the United States* remains the authoritative text in the field. "He was trying to reelect Carter."

If Volcker had been willing to help Carter, who had appointed him Fed chairman in 1979, then it stood to reason he'd listen to Reagan, who would have to reappoint him when his term expired in 1983. The White House staff only wanted the President to drop one or two hints. Reagan refused. "The only advice he ever gave Volcker," says Ed Meese, "was to use his best judgment and keep at it." By granting Volcker genuine independence, Reagan enabled the Fed chairman to succeed in wringing inflation out of the system. Reagan also established a precedent that has permitted the Federal Reserve to retain its independence to this day. "No other President would have stood by in 1981 and 1982 and said to the Fed, 'Keep doing what you're doing. You're doing the right thing,' " says Milton Friedman. "But Reagan understood that in order to break inflation he had to take a recession, and he had the courage to pay the political price."

> {T}here were always those who told us that taxes couldn't be cut until spending was reduced. Well, you know, we can lecture our children about extravagance until we run out of voice and breath. Or we can cure their extravagance by simply reducing their allowance.
>
> —Ronald Reagan, address to the nation on the economy,
> February 5, 1981

Our second scene: A 1982 meeting of the President's Economic Policy Advisory Board, a group of economists that included Milton Friedman, George Shultz, Alan Greenspan, Arthur Burns, and Arthur Laffer. Meeting in the Roosevelt Room, the Advisory Board considers a deal that chief of staff James Baker has made with Senate Finance Committee chairman Bob Dole and House speaker Tip O'Neill. If Reagan agrees to a tax hike, Dole and O'Neill have promised, they'll see to it that for every dollar the tax hike brings in Congress will cut three dollars in spending. As the President listens, he flexes his jaw, a sign that he's unhappy.

A word of background. During the campaign of 1980, Reagan had promised to cut taxes, not hike them. But he had also promised to boost defense spending and reduce the budget deficit. Then the recession had struck. When making good on all three of his campaign promises had become a mathematical impossibility,* Reagan had chosen to make good on only the first two, cutting taxes and boosting defense spending while permitting the deficits to mount. Soon it became clear that the administration would find itself presiding over the biggest peacetime deficits in American history.

Now, I myself was never able to get worked up over the deficits. I still can't. During his eight years in office Reagan added a total of $1.4 trillion to the federal debt. Yet what did he get in return? A military buildup that prompted the collapse of the Soviet Union and an economic expansion that raised incomes across the United States while creating some eighteen million new jobs—although many economists predicted the deficits would force interest rates up, dampening economic growth, interest rates instead fell, spurring economic growth. By the time Reagan left office, indeed, the value of assets in the United States—stocks, land, patent rights, and so on—had increased by $17 trillion, an amount twelve times bigger than the amount Reagan had borrowed. World peace plus a return of twelve to one. Not bad, I've always felt.

But just try telling any of that to Reagan's senior staff. James Baker, the chief of staff, Richard Darman, an assistant to the President, David Stockman, the director of the Office of Management and Budget, David Gergen, the assistant to the President for communications—all became convinced the deficits would do the economy irreversible harm. To bring the deficits under control, they had to bring Ronald Reagan under control, so

---

* By contracting the economy, the recession had reduced federal revenues. But by raising the unemployment rate, the recession had obligated the government to provide more unemployment benefits, increasing federal expenditures. Revenues down, expenditures up. Suddenly the administration found itself with a budget shortfall of tens of billions of dollars.

even before his tax cuts had taken effect they began trying to persuade the President to take them back. They leaked to the press, fanning alarm over the deficits, while working with Congress to produce tax hikes they hoped the President would feel compelled to sign. The deal Baker made with Dole and O'Neill represented part of this effort to convince Reagan to reverse himself.

To return to our scene: After the Advisory Board has listened to the terms of the deal, Arthur Burns, the former Fed chairman whom Reagan had named ambassador to West Germany, urges the President to take the deal. "Burns says, 'My best advice to you, Mr. President,' " Arthur Laffer recalls, " 'is that you should accept a revenue enhancer in order to get Congress to agree to these major cuts in spending.' "

"The President turns to Burns," Laffer continues. "Reagan says, 'You know, Arthur, I can't tell you how much I enjoy these Advisory Board meetings. But you know, I made a promise when I ran for office that I wouldn't raise taxes, and I intend to do all I can do to keep it. So every minute you spend in these meetings talking about a tax increase is a minute I don't get the pleasure of discussing something I might actually do.'

"And then," Laffer says, "the President leans over to Burns. He says, 'Never mention a tax increase in my presence again. Is that clear?' Just like that. Bam."

Reagan did in the end take back part of his tax cuts, reluctantly imposing a federal excise tax on gasoline, raising Social Security taxes, and repealing certain corporate tax breaks. He even signed the 1982 Tax Equity and Fiscal Responsibility Act, or TEFRA, a version of the deal that Baker made with Dole and O'Neill.* Was Reagan's rebuke of Arthur Burns therefore an

---

*TEFRA raised taxes by some $90 billion over the next three years. Those three dollars in spending cuts for every dollar in new revenues that Dole and O'Neill promised? Edwin

empty gesture? Hardly. Reagan found Congress, the press, and most of his own senior staff arrayed against him—as Laffer puts it, "He had everybody crapping on him. *Everybody.*" A practical politician, Reagan knew he would have to make concessions. Yet by resisting each concession so fiercely, Reagan was able to preserve by far the most important of his tax cuts, the across-the-board cuts in personal income tax rates. "The rate cuts were the alpha and omega of Reagan's whole program," Laffer says. "Why did the economy begin to boom in 1983? Because that was when the full rate cuts finally took effect. That boom would never have happened if Reagan hadn't fought the way he did."

> *America was born in the midst of a great revolution sparked by oppressive taxation. There was something about the American character—open, hard-working, and honest—that rebelled at the very thought of taxes that were not only heavy but unfair. Today the proud American character remains unchanged. But slowly and subtly, surrendering first to this political pressure and then to that, our system of taxation has turned into something completely foreign to our nature—something complicated, unfair, and, in a fundamental sense, un-American.*
>
> *Well, my friends, the time has come for a second American revolution.*
>
> —*Ronald Reagan, remarks in Oshkosh, Wisconsin,*
> *May 30, 1985*

The third scene takes place in the Oval Office as Richard Wirthlin, the President's pollster, meets the President in the spring of 1985. Reagan wants to pursue legislation that would

---

Meese has examined several studies analyzing just how much Dole and O'Neill actually delivered. "The estimates," Meese says, "range from sixty-nine cents down to zero."

simplify the tax code, remove millions of working poor from the tax roles, and cut personal income tax rates once again, lowering the top rate to 28 percent. Wirthlin wants to talk him out of it.

"There was a reason I felt discouraged," Wirthlin now says. "We'd been polling, and when we asked the open-ended question, 'What is the first thing that comes to mind when you hear the phrase, "tax reform"?' the majority of Americans said, 'Tax increases.' They opposed tax reform because they believed it would mean a tax increase. I never thought I'd have to say this about a project Ronald Reagan wanted to take on, but it looked like we were going to lose."

Wirthlin tells the President the polling numbers are against him, arguing that tax reform will at best prove "an uphill battle." Then he pushes Reagan to drop or modify his support for the legislation. Reagan pushes right back, refusing. "He certainly responded to my report soberly," Wirthlin says. "But it didn't in any way deter him."

Later that spring, Reagan began criss-crossing the country to campaign for tax reform. "When they heard about it from Reagan himself," Wirthlin says, "people decided they could trust tax reform. He was able to tap into their values. That man was amazing." The President continued to campaign for tax reform for more than a year, once even visiting Capitol Hill to lobby for the measure in person. In late 1986, after it finally passed both houses of Congress, Reagan signed the Tax Reform Act into law.

L ETTING VOLCKER do what Volcker wanted to do despite knowing many on his staff wished he wouldn't. Enacting deep tax cuts and then, in the face of intense political pressure, rolling back the tax cuts only grudgingly and in part while flatly refusing to compromise the most important cuts, the cuts in personal income tax rates. Campaigning for tax reform despite polling data showing he'd lose. When I thought about the way Reagan had enacted his economic program, an old gag from *Candid Camera* came to mind. In the gag, a camera was hidden in an elevator. When the elevator doors opened, the mark—an unsuspecting office worker—would get in, then do what anybody in an elevator would do, turn around to face the front.

At the next floor another one or two people would enter. Instead of turning around to face the front, however, these riders, in on the gag, would continue to face the back. The mark would glance at them, puzzled. At the next floor yet another couple of people would enter, and, again, instead of turning around to face the front, they'd face the back. The mark would grow nervous. He'd cast sidelong glances at his fellow riders. His face would begin to twitch. Finally he'd turn around to face the back himself.

I kept picturing the Gipper as the mark. He found himself riding an elevator in which economists, members of Congress and the press, and quite a few officials in his own administration, intent on their own economic policies, including higher taxes, faced not the front of the elevator, but the back. Yet Reagan himself refused to turn around, foiling a foolproof gag. It was this stubbornness, I saw, that enabled the President to enact his economic program. But what accounted for the stubbornness itself? What enabled Reagan to prove so impervious to supposedly expert opinion? To remain so unyielding despite all the political pressures he encountered? To pose the question yet again, How did Reagan do it?

There were two related answers. I figured one out right away. The other took a while.

—— *Hole After Hole* ——

ONCE YOU'D written a few economic speeches for Reagan, you'd have figured out the first answer yourself. Ronald Reagan just looked at economics differently from most people. When he took office, he stepped up to an economic model—a coherent, interlocking set of assumptions about the way economies work—that had dominated the Western world for at least four decades, then blithely turned it upside-down, replacing it with a model of his own. The model he overturned was that of Keynesian economics,* the model with which he replaced it, that of supply-side economics.

"If you look at the traditional Keynesian model," says economist Gary Becker, "you'll see that it pays attention to aggregates—aggregate demand, aggregate income, and things of that type."

The Keynesian model, in other words, concerns itself with big, abstract forces. It lumps together the demand for goods and services in the entire economy, calling the result "aggregate demand," and lumps together the income of every person in the entire economy, calling the result "aggregate income." Then it asks what government can do to control these colossal abstractions. Typically, the Keynesian model concludes that government must increase "aggregate income" by increasing "aggregate demand," which government must in turn accomplish by spending lots of money—and levying taxes to support such spending. The outcome? More government.

---

*Named after the British economist John Maynard Keynes, whose Great Depression treatise, *General Theory of Employment, Interest, and Money,* established the Keynesian school.

"Supply-side economics, by contrast," Becker says, "came along and asserted that maybe we should worry less about aggregates and pay more attention to the effect of incentives—especially taxes and regulation—on individuals."

Supply-side economics, in other words, sets aside big, abstract economic forces to concern itself instead with ordinary people. Typically, the supply-side model suggests that the economy needs not more government, but less. Cut taxes, reduce regulations, and control inflation, and people will go about their business with more than enough energy and ingenuity to keep the economy healthy.

Reagan didn't invent supply-side economics, of course. But he adopted it when it remained a fringe movement. (The very name "supply-side" was originally a term of derogation. Addressing a 1976 conference, Herbert Stein, who had served as chairman of President Nixon's Council of Economic Advisers, spoke mockingly of "supply-side fiscalists" who had the support of "maybe two" economists.) "I'd have lunch with Reagan at the Beverly Wilshire [Hotel] between 1976 and 1980," says Arthur Laffer, who tutored Reagan in supply-side economics. "It was just the two of us. He'd ask questions. I'd answer. By 1980, he had the whole model. Bingo. That was it."

Ronald Reagan, picking up an entire economic model over a few hotel lunches. Whereas Keynesian economics is the economics of professors, career politicians, bureaucrats, and others who wish to place their hands on the levers of government Reagan grasped, supply-side economics is the economics of ordinary people.

Which brings me to the second answer, the one that took me a while. Ronald Reagan understood the meaning of work.

Now, while I was still on the Vice President's staff, I assumed the President was lazy. Reagan, I saw, kept short hours. He spent his weekends at Camp David relaxing. Every summer he'd fly out to his California ranch to spend nearly the entire month of August on vacation. The most powerful man in the world looked like one of the most idle. Then, when I joined the President's speechwriting staff, I noticed something odd. In editing the drafts his speechwriters produced for him, the President was more diligent than the Vice President, not less.

Often the Vice President would look at a speech only when giving it was the next item on his schedule. When we were traveling, the routine was all but invariable. Aboard Air Force Two, I'd feel the nose of the aircraft tip downward, indicating that we'd begun our descent. With perhaps twenty minutes remaining before we landed and another half-hour before he'd deliver the speech, the Vice President would send back his personal assistant, Joe Hagin, to call me to the forward cabin. Sometimes the Vice President would hand me a set of edits. But often he'd simply place the stack of cards on which the speech was typed in the palm of one hand, then move his hand up and down as if holding a lead weight. "I like it," he'd say, "but it's a little heavy. See if you can cut it in half."

I'd return to the rear cabin, find the Vice President's secretary, Linda Casey, and begin dictating. We'd work aboard Air Force Two, continue working in a car in the motorcade, then finish the rewrite in the room that would have been set aside for us in the hotel, high school, or convention center in which the Vice President was about to appear. And then I'd scoop up the new pile of index cards to dash down one hallway after another, Secret Service agents directing me at each turn, to deliver the new text to Bush.

Mad dashes on the Reagan staff? Unthinkable. You'd hand in

your draft one afternoon, then receive the President's markups the next morning. Sometimes Reagan rewrote speeches extensively. Looking over the records in the Reagan Library recently, I found that he'd rewritten about a third of the 1983 "evil empire" speech. Sometimes he made very few changes. One morning I seated myself at my desk with a cup of coffee, prepared to devote a good hour to the President's changes on my most recent speech. No changes on the first page. None on the second. None on the third. As I flipped through the rest of the text, the thought crossed my mind that for once the Gipper had sent back a speech without even looking at it. Then I came to the last page. In the second-to-last line, the President had changed a single word.

Reagan never worked into the small hours of the morning like his predecessors Richard Nixon and Jimmy Carter or his successor Bill Clinton. But he worked steadily. When he left the Oval Office for the residence, flew to Camp David, or departed for his ranch, Jim Kuhn, the President's personal aide, informed me, Reagan always took with him speeches to edit, bills to sign, briefing materials to study, and letters to write. Reagan was capable of working so hard, Dana Rohrabacher, my colleague on the speechwriting staff, explained, that during the 1976 and 1980 presidential campaigns much of the staff had trouble keeping up with him.

"We'd be flying back to California after a campaign swing," Dana now says. "Everybody would be just pooped. Dead tired." Soon everyone would be sound asleep—with one exception. "Reagan would always be sitting up at the very front with the light on. He'd be going over notes, reading things, maybe writing out new cards for speeches. Sometimes I'd go up and sit next to him and we'd talk. That's really when I got to know him. You could always count on him to be awake while everybody else was out cold."

Late at night on a darkened plane, one light. Beneath it, Ronald Reagan. Working.

What impressed me most was the work Reagan performed at his ranch. Whenever he visited Rancho del Cielo, I learned, Reagan would spend hours clearing trails, pruning trees, chopping wood, hauling off weeds and scrub, and putting up fencing, digging hole after hole for fenceposts. Bill Wilson, a wealthy friend of Reagan's who owned a lemon ranch at the foot of the mountain, would have sent workers up the road to Rancho del Cielo anytime Reagan wanted. Assisted only by Barney Barnett and Dennis LeBlanc, former state policemen he had employed since he was governor, Reagan instead worked the place himself. Why? There could only have been one explanation. Reagan saw the work he performed at the ranch as an end in itself. He worked for the sake of the work.

"All the great saints looked at work the same way," Father Albacete told me over another dinner at Kentucky Fried Chicken. "The human being's greatest act—his freest, most intense, most human act—is offering himself up in worship. Properly understood, work is a form of worship. A man can offer up the weariness and difficulties of his labor. Or he can offer up the pleasures of solving problems or working with his hands. Work is the way human beings worship God by collaborating with him in the ongoing process of creation."

The meaning Reagan found in work, I saw, bore directly on his policies. If work wasn't just a way for people to meet their material needs but something like an act of worship, then government had better treat it with respect. Monetary policies that debased the unit of exchange? Regulations that stifled innovation? Tax rates that punished success? These represented more than errors. They represented affronts. Reagan pursued his economic policies with passion and determination because he was attempting not merely to nudge the gross domestic product up a couple of points, but to enable Americans to reap richer rewards and derive greater satisfaction from the activity in which ordinary people found themselves engaged most of their lives. He was defending the dignity of work.

——*A Step Ahead*——

ONE DAY Josh Gilder, stuck on a speech he was writing, sauntered into my office, flopped down on my sofa, and began shooting the breeze. That was fine with me. I was stuck on a speech, too. I put my feet up on my desk and began shooting the breeze right back. Our conversations were always free-form, wandering from topic to topic, but sooner or later they often settled, as did this one, on our work. I offered up some of the usual speechwriters' bellyaching about how hard writing was, which was true. All of us did find writing hard. Then I began on another staple complaint, grousing about how we speechwriters were underpaid, underappreciated, and looked down upon, which was true, too. Almost everyone in the White House except Ronald Reagan himself treated the speechwriters as if we were merely some sort of steno pool. Writing, they assumed, was only a small step above typing.

"Yeah, I know," Josh replied. "But when you're writing, don't you feel as though you're working on your soul?"

Often a step ahead of me, Josh had already made Ronald Reagan's ethic his own. The right way to look at work, I recognized in that moment, was to see it as soulcraft.

I still recall the peace of mind I felt when I returned to my writing after Josh left. It would be nice if I ever ended up in a job that felt like an adornment, I still thought, and even nicer if someday I found myself performing work for which I was paid a lot of money. But the value of work lay not in the kind of social position it provided or, beyond the obvious need to cover the bills, in the amount it paid. The value of work lay in allowing you to develop your talents and build up your character. Money? I'd never turn up my nose at it. But collaborating in the process of creation—that was the point.

Looking at work in this way, I realized, meant I could stop wasting my time on daydreams. A captain of industry? Me? Ridiculous. In considering my career, I'd have to ask what talents I actually possessed, not what lifestyle looked attractive. In the meantime, I could settle down and derive a lot more satisfaction from my job at the White House. Who would have guessed I could write speeches? Yet as things had turned out, I could. A talent for turning phrases was hardly akin to a talent for science or the fine arts, needless to say. But it was a talent all the same, and I decided to make the most of it. Soon I developed a habit I still have today. When my job felt like drudgery, I'd picture Reagan at the ranch, driving a posthole digger into the ground again and again. Do your work, I'd remind myself. It's good for the soul.

——— *Three* ———

HOW TO ACT

Life Is a Drama. Do *Something.*

JOURNAL ENTRY, MAY 1990:

When I stopped by to see Reagan in his Los Angeles office today, I mentioned that I was in town to peddle a treatment for a television show. He smiled, then put his head to one side and gave it that famous little shake. "Well," he said, "don't let them do to you what they tried to do to me."

When he first arrived in Hollywood, Reagan said, he attempted to persuade his employer, Warner Brothers, to produce a movie about Knute Rockne, the great Notre Dame football coach, and George Gipp, the legendary halfback who died while still in his twenties. Reagan wanted to play Gipp himself. The studio displayed no interest. Then one day Reagan noticed an item in Variety, *the show business newspaper, announcing that Warner Brothers was about to begin production of* Knute Rockne—All American.

"Well," Reagan continued, "I marched right over to let the producer

know I was going to be his George Gipp. He told me I wasn't big enough. I said, 'Would it surprise you to learn that right now I weigh five pounds more than Gipp did when he played for Notre Dame?' That got his attention, but I still hadn't convinced him. So I drove home, found a yearbook photo of myself in my college football uniform, drove back to the studio, and paid a second call on the producer. When the producer saw that yearbook photo, he arranged for me to test for the part the very next morning. And the day after that he called to say, 'Reagan, you're playing the Gipper.' "

Even as a young actor, I realized afterward, Reagan had been completely Reagan. He'd imagined himself in the role of Gipp—and then he'd gone for it.

WHEN I STARTED WORKING at the White House, I had the idea that after completing the background reading I ought to be able to outline a speech in minute detail before I began writing it. I never could. I'd make a few pages of notes, find myself unable to put them in order, and then, feeling frustrated, give up. My inability to compose outlines produced two ill effects. It wasted a lot of time. And it made me feel inadequate, convinced that although my speeches usually turned out all right my technique was somehow deficient.

This small problem was related to a big problem. I couldn't figure out how much life itself was supposed to be like writing a speech. Should I be able to outline my future, setting out detailed goals? One of my best friends happened to be attending business school. In his job interviews, he told me, he was always being asked to describe his five-year plan.

"Your five-year plan for what?" I asked.

"For my life."

"Oh, come on. They expect you to have a five-year plan for your *life?*"

"They'd consider me unprofessional if I didn't."

So far all the good things in my own life had just happened to me. I'd attended Dartmouth after a high school guidance counselor had taken me aside to suggest that I apply to a school in the Ivy League. I'd attended Oxford after a Dartmouth professor who was a graduate of Oxford himself made a few telephone calls to his old college on my behalf. And I'd gotten a job at the White House—well, you've already seen how that came about. It looked to me as though I was doing all right by just sitting back and letting life happen. On the other hand, even I could see I'd been lucky—and that my luck couldn't last forever. But what was I supposed to do? Start drawing up five-year life plans when I couldn't complete a speech outline?

Ronald Reagan solved both problems. To explain, I need to show you how he won the Cold War. Yes, I know. From my problems to the Cold War. But once again, bear with me here.

—— *The Empty Photo* ——

WE'VE ALREADY looked at before-and-after pictures of the American economy. Now let's look at before-and-after pictures of our principal adversary during the Cold War, the Soviet Union.

In the picture that portrays our adversary before Ronald Reagan took office, the Soviet Union appears self-confident, swaggering, belligerent. And why not? Between World War II and the late 1970s, the Soviet Union assembled the biggest army on earth, transformed its coastal defense force into a vast blue-water navy, and amassed an arsenal of thousands of nuclear weapons. It crushed uprisings in East Germany in 1953, Hungary in 1956, and Czechoslovakia in 1968. It armed and funded Communist insurgencies in Vietnam, North Korea, and more than half a dozen nations in Central America, Latin America, and Africa. In 1977 it began deploying intermediate-range nuclear missiles that subjected Western Europe to a direct nuclear threat, and then, in 1979, it engaged in a major military adventure, invading Afghanistan.

The after picture? A snapshot of nothing. On December 25, 1991, the Soviet Union was officially dissolved.

No tank battles on the plains of Central Europe. No nuclear exchanges. In short, no World War III. Yet in less than a decade, the Soviet Union, an expanding empire at the beginning of the period, had become defunct. In the words of former British prime minister Margaret Thatcher, "Ronald Reagan won the Cold War without firing a shot."

How did Reagan do it?

—— *We Win and They Lose* ——

HERE AGAIN, I realized, there were two answers. The first? Reagan's capacity for imagination. I came to appreciate this answer during a bull session one evening in the office of chief speechwriter Tony Dolan.

"What's distinctive about acting?" Tony asked, waggling an unlit cigar up and down in his mouth as he, Josh, and I talked about Reagan's career in motion pictures. "I'll tell you what. Actors get used to the idea of alternative endings."

Reagan became an actor at a time when Hollywood made

dozens of motion pictures a month. "The studios were just crank-ing those suckers out," Tony said. This enormous volume of out-put led to chaotic working conditions, and screenwriters in particular often fell behind. When they did, actors shot scenes without knowing what came next—or even wrote scenes them-selves. "Reagan was supposed to be pretty good at knocking out a scene or two right on the set," Tony said. "Anyway, if you act under those conditions for a while you'll get used to the idea that there's nothing preordained about any script. Scenes can be rewritten. New endings can be added."

"And this is relevant to—what, exactly?" Josh asked. Tony proved so easy to annoy that we could never resist the temptation.

"Don't rush me, not that either of you sacks of garbage has anywhere to go tonight," Tony said. He chewed his cigar for a moment, shaking his head.

"The thing about Reagan," Tony continued, "is that he was able to make the leap from acting to reality. He understands open-endedness and contingency. He sees that life is a drama in which a lot of the scenes still haven't been written. And recog-nizing the open-endedness of life makes Reagan a lot more un-usual than you might think."

Although nobody could tell you when he climbed out of bed in the morning just how his day would unfold, Tony argued, a lot of people in public life had been behaving for a couple of decades now as though they knew how the entire century would unfold. The Western world, they assumed, was in decline, destined to lose to Communism. Even conservatives had proven gloomy. In his 1965 autobiography, *Witness*, for example, Whittaker Chambers, the one-time spy who became a prominent conserva-tive, wrote that when he broke with the Communist Party he felt certain he was "leaving the winning world for the losing world."

As late as 1985, during the Reagan administration itself, the French conservative Jean-François Revel published a grim little book entitled *How Democracies Perish*.

"When Reagan says he doesn't understand how anybody could handle the presidency without having been an actor," Tony said, referring to a remark the President often made, "I figure he has a couple of things in mind. One is the obvious stuff. A President needs to know how to stand in front of a camera and deliver his lines. But Reagan is also saying that a President needs an actor's imagination. He has to be able to size up a script. Chambers, Revel—all those guys were playing a tragedy. But Reagan can imagine a post-Soviet world—he can really *see* it. So he tossed out the old script to write a new script of his own. What the Gipper likes is happy endings, not tragedies."

Tony chewed his cigar. Then he looked at us. "The future of civilization. Is that relevant enough for you?"

N OT LONG AFTER that bull session I heard a story that illustrated Tony's point. Since then, I've confirmed the story with the source, Reagan's first national security adviser, Richard Allen.

JOURNAL ENTRY, SEPTEMBER 2002:

"In early January 1977 I spent a morning and afternoon talking with Reagan about foreign and defense policies," Richard Allen writes in an e-mail he sent me today. "Reagan asked if I'd like to hear his theory concerning the Cold War and the Soviets. I allowed as how I certainly would. Reagan said, 'Some people think I'm simplistic, but there's a difference between being simplistic and being simple. My theory about the Cold War is that we win and they lose. What do you think about that?'

"I was flabbergasted," Allen writes. "I'd worked for Nixon and Goldwater and many others, and I'd heard a lot about Kissinger's policy of détente and about the need to 'manage the Cold War,' but never did I hear a leading politician put the goal so starkly.

" 'Governor,' I asked, 'do you mean that?'

"Reagan said, 'Of course I mean it. I just said it.' "

WE WIN and they lose." Simple as it was, I recognized, the President's strategy had required him to engage in an audacious act of imagination. He'd had to be able to envision a United States willing to marshal its military, economic, and technological superiority as never before, subjecting the Soviet Union to pressures that were new, intense, and sustained. Reagan was winning the Cold War, in other words, for the very reason Tony Dolan had named. The President was able to *see* a post-Soviet world.

Reagan's capacity for imagination—that was the first answer. The second? This takes us back to Tony Dolan.

—— *How to Act* ——

THIS BULL SESSION took place in my office. Josh and I were shooting the breeze when Tony shoved open my door, flopped down in one of my armchairs, put his feet, which were, as usual, in cowboy boots, on my coffee table, and began telling us about a book he'd read, *In Search of Excellence.*

"A *business* book?" Josh asked.

"Yeah, well, every so often even the suits figure something out."

On the *New York Times* bestseller list for more than three years, *In Search of Excellence* surveyed dozens of the best-run corporations in the country, examining the characteristics they shared. One chapter had made a particular impression on Tony. Titled "A Bias for Action," it described how companies ranging from Hewlett Packard to Procter & Gamble to Bloomingdale's all responded when they encountered problems. They didn't waste time on feasibility studies or committee meetings. They *did* something. To quote the chapter itself:

> "Do it, fix it, try it," is our favorite axiom. . . . Getting on with it, especially in the face of complexity, does simply come down to trying something. . . . *The most important and visible outcropping of the action bias in the excellent companies is their willingness to try things out, to experiment.*

Tony had not merely read this chapter. He had found a place for it in his philosophy of life. "What the two of you would never know if I didn't tell you," Tony said, "is that business has ontological value. What does a businessman do? He takes action. And in the very act of acting—if you'll forgive me for sounding like St. Thomas Aquinas, not that either of you has ever even heard of the man, let alone read him in Latin, which I have, by the way— in the very act of acting, the businessman demonstrates mastery over the created order."

Consciously or not, Tony maintained, the businessman was a believer. The universe was an orderly entity and humans had a settled place within it—this was the businessman's implicit credo. He would never possess a detailed knowledge of all the laws of mathematics, economics, and psychology that governed the marketplace, but the businessman never doubted that such laws existed. If only he tried enough new products, manufacturing techniques, and so on, he understood, he'd achieve his objectives.

"It's like the command in Genesis," Tony said. " 'Have dominion over every living thing.' That's what businessmen do. They go around having dominion."

To this point Josh and I had been following Tony. Now he threw us.

"Reagan is like that," Tony said, rolling his unlit cigar from one corner of his mouth to the other. "He's like a really good businessman."

Reagan? A businessman? In his long life Ronald Reagan had been a radio announcer, movie star, union leader, television personality, and politician. Business seemed the one field for which he had never demonstrated any aptitude. Josh and I exchanged glances.

"Stop looking at each other like that," Tony said. "I haven't lost my mind yet, although if I have to keep working with you two it could be any day now. What I'm saying is that Reagan has a bias for action."

Like the businessman, Tony maintained, Reagan was a believer. He would never possess a detailed knowledge of all the laws that governed the lives of nations, but Reagan never doubted that such laws existed. "We respect human liberty and the Soviets don't," Tony said. "That means our society is in fundamental harmony with 'the laws of nature and of Nature's God'—that's the Declaration of Independence, by the way, not that either of you two has ever read the document—and theirs isn't. Reagan knows that. He sees that their system goes against the grain of reality.

"Reagan may not know exactly what it will take to bring the Soviets down," Tony continued, "but he's willing to experiment.

Rebuild the military. Kick the Cubans out of Grenada. Support the freedom fighters in Afghanistan and Nicaragua. Crank up a little research on Star Wars."

Tony chomped his cigar for a moment, letting his little discourse sink in. "A bias for action," he said. "Have you two got it now, or was I talking too fast?"

From a business book to ontology to the Cold War. It was, as it always was with Tony, a wonderful show, maybe the best in the White House.

The thing is, Tony was right.

—— *Comrade, Tell Me How* ——

OVER AND OVER in his dealings with the Soviets, I realized after that bull session, Reagan had demonstrated just the characteristic Tony claimed he possessed: a bias for action. Consider a few of the President's most famous lines, statements that we speechwriters repeated among ourselves admiringly. When in September 1981 arms negotiator Paul Nitze, exasperated, asked how he was supposed to present the President's new bargaining position to the Soviets—Reagan had tossed out Jimmy Carter's "two-track option," which would have left in place many of the intermediate-range nuclear force missiles, or INFs, that the Soviets had deployed in Eastern Europe, for a new "zero option," which would require the Soviets to dismantle their INFs—the President replied with a smile, "Well, Paul, you just tell the Soviets you're working for one tough son of a bitch." When in June 1982 the senior members of the administration gathered in the Cabinet Room to advise the President against economic sanctions intended to halt construction of a Soviet natural gas pipeline, the President said, "Well, they can build their damn pipeline." Looks of relief appeared around the table. Then he slammed down his fist. "But not with *our* technology." And then he stood and left the room. And when in October 1983 secretary of state George Shultz and national security adviser Robert "Bud" McFarlane telephoned Reagan in the middle of the night, informing him that six nations in the Caribbean had asked the United States to intervene in Grenada—infighting in the Marxist clique that ruled the island had produced violence and instability— Reagan, conscious of the hundreds of Soviet-backed Cuban troops on Grenada, thought for a moment, then spoke two words: "Do it."

Now, I'd always seen these famous lines as just that, lines— statements that demonstrated Reagan's feel for the dramatic. The President couldn't have played those scenes any better, I'd thought, if we speechwriters had scripted them ourselves. Now, though, I considered the lines in a new way, examining them instead as statements that demonstrated Reagan's capacity for leadership. Had Reagan moved slowly and cautiously? Had he played for time by asking for feasibility studies? Hardly. Reagan had *acted.*

The best illustration of Reagan's bias for action? That was easy: the program that more than any other belonged to him alone, the Strategic Defense Initiative, also known as Star Wars. Controversial even today, SDI is worth taking a moment to consider.

> *Let me share with you a vision of the future which offers hope.*
> *It is that we embark on a program to counter the awesome*
> *Soviet missile threat with measures that are defensive. Let us*
> *turn to the very strengths in technology that spawned our great*
> *industrial base and that have given us the quality of life we*
> *enjoy today.*

76

*What if free people could live secure in the knowledge that
their security did not rest upon the threat of instant U.S.
retaliation to deter a Soviet attack—that we could intercept
and destroy strategic ballistic missiles before they reached our
own soil or that of our allies?*

*I know this is a formidable technical task, one that may
not be accomplished before the end of this century. Yet current
technology has attained a level of sophistication where it's
reasonable for us to begin this effort . . .*

*My fellow Americans, tonight we're launching an effort
which holds the promise of changing the course of human
history.*

—*Ronald Reagan, address to the nation, March 23, 1983*

Though I was still on the Vice President's staff when Reagan
announced the Strategic Defense Initiative, I'd become friends
with a couple of the President's speechwriters—my office was on
the second floor of the Old Executive Office Building, directly
above theirs, and when I had a moment it was easy enough to slip
downstairs—so I'd known all about the big defense speech the
President intended to deliver. Or so I'd thought, anyway. A new
program to counter the Soviet missile threat? Where, I found myself thinking as I listened to the speech over pizza in my apartment, had *that* come from? When I got to my office the following
morning I no more than tossed my briefcase onto my desk before
running downstairs.

"Did you know anything about the announcement in the
speech last night?" I asked Tony Dolan. "All that stuff about
knocking ballistic missiles out of the sky before they can hit us?"

"I know you sleep better at night believing that Ronald
Reagan never utters a word without consulting me," Tony

replied, chewing a cigar, "but I'm afraid I have to disappoint you."

Dana Rohrabacher joined us, explaining that he himself had only learned about the announcement late the previous evening. "Somebody came downstairs from the National Security Council with this big insert to put in the President's address," Dana said. "I can tell you one thing, though." He grinned. "If we're all running around trying to figure out what happened, just picture the scene in the Kremlin."

Between reports in the press and what we heard in the White House, over the next couple of weeks we were able to piece events together.

Six weeks before, the President had lunched in the Cabinet Room with the joint chiefs of staff. Conversation had ranged over a number of topics—Reagan had made it clear that his guests should feel free to discuss whatever was on their minds—and Admiral James Watkins, the chief of naval operations, had mentioned that he'd always felt uncomfortable with "mutually assured destruction," or MAD, the doctrine on which American nuclear policy had been based ever since the Soviet Union developed nuclear weapons of its own. The United States could never prevent the USSR from destroying most of North America, MAD maintained. What the United States could do was make the Soviets a promise: If they launched a nuclear attack on us, we'd retaliate with a nuclear attack on them. Two continents wiped out instead of one.

"We should defend the American people, not avenge them," Watkins had said. He and his colleagues had then explained that a certain amount of theoretical work had been done on systems intended to destroy incoming nuclear missiles before they reached their targets. Reagan had responded by asking each of the chiefs what he thought of such a system. Although the discussion had covered the costs and scope of a system in only the most general terms, each chief had expressed his support in principle for a strategic defense.

What the chiefs may not have known was that the President knew a lot about the subject already. Ever since serving as governor of California, Reagan had been receiving briefings from Edward Teller, the physicist who participated in the Manhattan Project during World War II, then, during the 1950s, helped develop the hydrogen bomb. Associate director since 1960 of the Lawrence Livermore National Laboratory, one of the nation's three nuclear research centers, Teller had long believed that a strategic defense would prove practicable—and had told Reagan just that.

After lunch with the joint chiefs, the President had asked Judge William P. Clark, then his national security adviser, to begin work on a statement announcing research on strategic defenses. Clark had instructed Bud McFarlane, his deputy, to compose a draft, and the President's science adviser, G. A. "Jay" Keyworth, had then revised McFarlane's draft. The President, Clark, McFarlane, Keyworth, and a few of their aides—you could have counted the number of people privy to the most important innovation in nuclear policy on two hands and a foot. (And one of the tiny number, Bud McFarlane, apparently believed until the very last moment that the announcement was merely an exercise. "Bud drew up a memo the night before the speech laying out all the reasons the President shouldn't give it," Judge Clark now says. "He twisted my arm to get me to delay the President. Of course I wasn't about to do that.") The Cabinet had been informed of the announcement just forty-eight hours beforehand. And then, six weeks after the idea had occurred to him over lunch, Ronald Reagan had gone on the air to turn nuclear policy upside down.

ONCE THE Strategic Defense Initiative had gotten underway, Josh Gilder and I used to amuse ourselves by railing against the Union of Concerned Scientists, which was always deriding SDI as a dangerous fantasy, while trying to imagine how SDI must have looked to the apparatchiks in Moscow.

"So we've got a bunch of scientists saying that building a strategic defense system is going to be a fantastically complicated technical problem," Josh would say as we lounged in my office. "Of *course* it's going to be a fantastically complicated technical problem. Who ever doubted that? But now that we're spending a few billion dollars a year working on the problem, guess what? We're learning tons of new stuff about ballistics, computer science, lasers, rocket propulsion, and about five dozen other disciplines I don't even know the names of. And suddenly the Soviets are looking at a United States that's willing to make use of two strengths they can't even begin to match, our economy—it takes a big, vibrant economy to support an effort like this—and our capacity for technological innovation.

"Picture the poor slob of a KGB agent who's assigned to infiltrate the Silicon Valley companies that are producing technology we might use in SDI," Josh would continue. "The dynamism of the high-tech sector just totally overwhelms him. When he gets back to Moscow to report to Gorbachev, you know what he says? 'Comrade Chairman, in just one month in Silicon Valley I saw ten high-tech companies go bust and twenty high-tech companies get started. How, comrade, tell me how—how can I infiltrate such an industry?' The Soviets may outnumber us in conventional forces and nuclear warheads, but now *they're* the ones who aren't sleeping so well at night. If the members of the Union of Concerned Scientists are so bright, why can't they see *that?*"

Why indeed.

The Strategic Defense Initiative worked. Not, of course, that Reagan or his successors ever actually deployed a strategic defense.* Yet as a counterweight to "the awesome Soviet missile threat," to use the President's words, SDI succeeded even more completely than many who supported it had dared to hope. "We never believed in the umbrella," Admiral James Watkins, now retired, has said of the views of the joint chiefs of staff at the time they advised Reagan. "What we said is that if you confuse the Soviets [with a system that might destroy at least some incoming missiles] there would never be a first strike." SDI went the joint chiefs of staff one better. It confused the Soviets so thoroughly that they simply threw up their hands in surrender.

Not that they didn't at first resist. At the 1986 summit in Reykjavik, Iceland, Mikhail Gorbachev offered President Reagan dramatic reductions in nuclear weapons—but only if the United States would agree to limit SDI to laboratory testing, forgoing for at least a decade any experiments on firing ranges or in space itself. In effect, Gorbachev argued, the United States would have to promise to pursue SDI only in theory, not in practice. Reagan's response? Although the Soviets had made it clear they were willing to spend an additional day in Reykjavik, Reagan closed his briefing book, stood up, and said, "The meeting is over."

In 1992, the year after the Soviet Union was dissolved, I attended a dinner at which former secretary of state Henry

---

* More's the pity. "It's a sin and a crime," former secretary of state George Shultz now says, "that by this time we haven't learned how to defend ourselves against [at least] a light missile attack."

Kissinger described a trip he had just made to Russia. Speaking to high officials in the government and military, Kissinger had asked each to name the critical factor in the demise of the USSR. "Almost without exception," Kissinger said, "they named SDI."

"The Soviets may have overestimated our technical capacity," Kissinger now says. "On the other hand, we didn't have to build a complete version of SDI to make their calculations difficult. If the Soviets no longer knew how many missiles would get through, then they might have had to launch hundreds more to have had a chance of success." Hundreds more? But the Soviets could never have *afforded* hundreds more. SDI had thus threatened to deprive the Soviet Union of its capacity to deliver an over-whelming nuclear strike—and without that capacity, the USSR would have looked a lot less like a superpower and a lot more like one more poor, backward nation. "You can see," says Kissinger, "why SDI had them rattled."

" 'Do it, fix it, try it,' is our favorite axiom," to quote *In Search of Excellence* once again. "Getting on with it, especially in the face of complexity, does simply come down to trying something." Ronald Reagan knew how to try something.

—— *The Drama* ——

CONTINGENT, OPEN-ENDED, unpredictable—life is a drama. Use your imagination, then *act*. Like all the lessons I learned from Ronald Reagan, this one proved scalable. Although I saw it operating in the life of one of the largest figures of the twentieth century, I was able to adapt it to the modest dimensions of my own life.

This brings me back to the problems with which I began this chapter. Elaborate speech outlines? They bore the same relationship to speechwriting, I saw, that feasibility studies and committee meetings often bore to the practice of business or government. They became an end in themselves. So instead of trying to compose an outline for each speech, I began simply picturing Reagan in front of the audience, then imagining what he might like to say. Afterward I'd jot down the barest of notes—only a few lines, even for a big speech—and start writing. I'd have to do a lot of rewriting as I worked, of course, but now I found myself confronting the substance of the speech at once. And when I had an idea I wasn't sure would work, I'd no longer stare at the ceiling, wondering whether to put it into the outline. I'd *write* it. Maybe it worked. Maybe it didn't. But once I saw it in print, I knew. The new approach saved time—well, some time, anyway. And there were days when it enabled me to feel almost as if I knew what I was doing.

My second problem, you'll recall, was whether to plan my life or just let it happen. After observing the fortieth chief executive for a while, I saw that neither made much sense. Plan life? How? One moment you'd be sound asleep. The next you'd have the secretary of state and national security adviser on the line, asking whether you'd like to mount an invasion of an island in the Caribbean. But just sit back and let life happen? When had Reagan ever behaved like that? He'd always proven engaged and determined, the kind of man who in a matter of weeks could turn an interesting idea he'd picked up at lunch into national policy. I decided I'd quit worrying about planning the unplannable, but that I'd also quit dreaming about relying on dumb luck. Instead I'd imagine a simple idea or vision for my life, then improvise, acting on my vision as opportunities arose.

For a time, I proved foolish enough to see myself as a banker, and after my first year at business school I took a summer job with Dillon Read, an investment bank in Manhattan. The work bored me—heaven help me, because that job probably represented the clearest shot I'll ever have at becoming rich, but the work bored me—and I knew within a couple of days that I'd made a mistake. But that was, in a way, the whole point. I'd had an idea, then done something about it. I'd *acted*. And as soon as I'd realized my mistake, I was able to start thinking through my vision or idea for my life all over again. Taking action, I learned, forces you to confront reality—and reality has a way of letting you know when you're wrong.

I had to make several more foolish moves—for a time I even became a junior television executive, just as we have seen—before deciding I'd go right back to doing what I did at the White House, spending most of my time as a writer. But each time I made a mistake, I discovered something new about my interests and abilities. I'd thought following Reagan's example would make my life easy. It didn't. But it proved the right approach all the same. It taught me just what I needed to learn.

——— *Four* ———

TEAR DOWN
THIS WALL

Words Matter

JOURNAL ENTRY, BERLIN, MAY 2002:

Although I only needed to be driven to my hotel, when I climbed into the cab after landing at Tegel Airport this morning I realized I could tell the driver to take me anywhere I wanted. To East Berlin? To East Berlin. Or to Potsdam, Leipzig, or Dresden. For that matter, if I'd had the fare I could have told the driver to head east until we hit Moscow.

The wall is gone.

——— *Bugle Boy* ———

THIS MAY SOUND like an odd admission coming from a former speechwriter, but when I first joined the President's staff I sometimes wondered whether his speeches really mattered.

It didn't help, I suppose, that as the youngest member of the staff I got stuck early on with a lot of assignments the other speech-writers wanted to avoid. Looking over the records in the Reagan Library recently, I found that during my first year I wrote more than 150 speeches, radio talks, toasts, and sets of talking points, at least half of them forgettable. Consider, for example, the remarks I composed for the 1984 Rose Garden ceremony to mark the fiftieth anniversary of the Migratory Bird Hunting and Conservation Stamp, a name I am not making up. Purchased each year by hunters, the Duck Stamp, as it was also known, produced revenues used to preserve waterfowl habitat. The ceremony itself may have held some importance—for all I know, it helped the President carry states that contained high concentrations of hunters on Election Day later that year. But the words I composed? Let's put it this way: You won't find them in *Bartlett's Familiar Quotations*.

As I thought about his speeches those first months, I found myself pondering Reagan's entire political career. Since I'd spent a year and a half working for Vice President Bush before joining the President's staff, contrasts between Bush and Reagan came readily to mind. George Bush, I recognized, spent a lot of his time tending to what I came to think of as his network. Aboard Air Force Two after giving a speech, the Vice President would sit down, pull out a stack of engraved note cards, then compose thank-you notes to all the officeholders, businessmen, and Republican officials who had sat with him at the head table. When he got done, he'd begin placing telephone calls. Listening to him, you'd find yourself wondering whether there was a single member of either the old corporate establishment in the Northeast or the new oil, gas, and real estate establishment of the Southwest with whom he *wasn't* on a first-name basis. Last the

Vice President might place a few calls to his family, getting in touch with siblings and offspring throughout the country. His brothers, Prescott, Johnny, and Bucky, lived, respectively, in Connecticut, New York, and Missouri, while his sister, Nancy, lived in Massachusetts. His sons, George, Jeb, Neil, and Marvin, lived, respectively, in Texas, Florida, Colorado, and Virginia, while his daughter, Doro, lived in Maine. Six degrees of separation? For Bush, that would have been about four degrees too many. Virtually every person of consequence in the entire country was either a personal acquaintance of Bush himself, of a member of his family, or of one of his friends.

A lot of politicians, I saw, possessed networks like that of George Bush. Richard Nixon was famous for his ability to greet Republican officeholders and party officials in every region of the country by name. Lyndon Johnson spent a lifetime performing political favors while committing to memory the identities of each of those on whom he had bestowed his largesse. John Kennedy's network included his father's business associates, his many siblings and their friends, and the intellectual establishment of the Eastern Seaboard, with which he ingratiated himself by bringing prominent academics such as John Kenneth Galbraith and Arthur Schlesinger Jr. into his administration. Franklin Roosevelt used to amuse guests by inviting them to draw a line across a map of the United States, then identifying the leading Democrats in each county through which the line passed. Franklin Roosevelt, John Kennedy, Lyndon Johnson, Richard Nixon, George Bush—each possessed his own enormous network.

In this regard, I realized, Ronald Reagan proved utterly atypical. Although, like Nixon and Johnson, he had come from a family of no social or business standing, unlike Nixon and Johnson he had never attempted to compensate for his lack of family connections by working with special zeal to establish a network from scratch. When a member of his staff asked him to place a telephone call to an officeholder or supporter, Reagan was always happy enough to comply. But pick up the telephone on his own? "Very rarely did you ever see him call anyone just to say 'howdy,' " Michael Reagan says. And although he wrote a lot of letters, Reagan displayed the same pattern when he became a politician that he had displayed as a movie star, addressing most of his correspondence to ordinary citizens, not to people of political consequence. He enjoyed keeping in touch with his fans, but he never demonstrated much enthusiasm for cultivating the powerful.

How had Reagan done it? What had he possessed that had enabled him to capture the highest office in the nation without an enormous network of the kind on which so many other politicians depended? Maybe the reason it took me a while to see the answer was that as a speechwriter I was so close to it. I kept thinking Reagan must have developed some political trick or technique that he kept tucked away, reserving it for his own use alone. But the answer lay in plain sight. What Reagan had possessed, I finally recognized, was words. Just words.

"Not too long ago two friends of mine were talking to a Cuban refugee, a businessman who had escaped from Castro," Reagan said in the televised 1964 speech on behalf of Barry Goldwater that established him as a national figure. "And in the midst of his story one of my friends turned to the other and said, 'We don't know how lucky we are.' "

And the Cuban stopped and said, "How lucky you are! I had someplace to escape to." In that sentence he told us the entire story. If we lose freedom here, there is no place to escape to. This is the last stand on earth. And this idea that government is beholden to the people, that it has no other source of power except the sovereign people, is still the newest and most unique idea in all the long history of man's relation to man. . . .

You and I have a rendezvous with destiny. We will preserve for our children this, the last best hope of man on earth, or we will sentence them to take the first step into a thousand years of darkness. If we fail, at least let our children and our children's children say of us we justified our brief moment here. We did all that could be done.

It might have been almost twenty years since Reagan delivered that speech by the time I read it in the White House, but I still found it powerful. It moved me. It rang in my ears. I know this isn't exactly an original simile, but I was never able to think of a better one. The words sounded like the call of a trumpet. In the old war movies I used to watch on Saturday afternoons when I was a kid—the cavalry fighting the Indians, the Union soldiers the Confederates, or the British the Zulus—it was always the same. The battlefield would begin to dissolve in confusion, the good guys finding themselves forced steadily back. Then a bugle or trumpet would sound, piercing the din. The good guys would cheer, rally, and then press on to victory. Reagan had no need for an enormous network because his speeches alone proved capable of rallying the American people to his cause. (While working on this book, my assistant learned that her mother found Reagan's 1964 speech so powerful that she went door to door in her Pasadena neighborhood, giving a copy of the speech to each of her neighbors.) Not all his speeches produced the same trumpetlike effect—not, obviously, his remarks on behalf of the Duck Stamp. But did Reagan's speeches matter? Enough, I realized, to make him President.

And enough, as I realized still later, to change the world.

—— *Four Blasts* ——

YOU CAN TRACE the whole story of Ronald Reagan's victory in the Cold War, I now see, by looking at just four of his speeches. In the first of the four, delivered to the British Parliament on June 8, 1982, Reagan laid out his strategy for exploiting Soviet economic weakness.

> In an ironic sense Karl Marx was right. We are witnessing today a great revolutionary crisis, a crisis where the demands of the economic order are conflicting directly with those of the political order. But the crisis is happening not in the free, non-Marxist West, but in the home of Marxist-Leninism, the Soviet Union. . . .
>
> The dimensions of this failure are astounding: A country which employs one-fifth of its population in agriculture is unable to feed its own people. Were it not for the private sector, the tiny private sector tolerated in Soviet agriculture, the country might be on the brink of famine. . . . Overcentralized, with little or no incentives, year after year the Soviet system pours its best resources into the making of instruments of destruction. The constant shrinkage of economic growth combined with the growth of military production is putting a heavy strain on the Soviet people. What we see here is a political struc-

ture that no longer corresponds to its economic base, a society where productive forces are hampered by political ones. . . . [T]he march of freedom and democracy . . . will leave Marxism-Leninism on the ash heap of history. . . .

The Soviet Union facing a crisis? Communism destined for the ash heap of history? Still trying to complete my doomed novel, I was in London the day Reagan delivered this speech. It represented such an affront to all the old, settled notions of coexistence that none of my English friends could quite believe their ears.

Having laid out the economic case against Communism, Reagan laid out the moral case, speaking on March 8, 1983, to the National Association of Evangelicals in Orlando, Florida.

Yes, let us pray for the salvation of all of those who live in that totalitarian darkness—pray they will discover the joy of knowing God. But until they do, let us be aware that while they preach the supremacy of the state, declare its omnipotence over individual man, and predict its eventual domination of all peoples on the earth, they are the focus of evil in the modern world. . . .

You know, I've always believed that old Screwtape [the demon in C.S. Lewis's book *The Screwtape Letters*, Screwtape was always trying to find new ways of corrupting human beings] reserved his best efforts for those of you in the church. So, in your discussions of the nuclear freeze proposals, I urge you to beware the temptation of pride—the temptation of blithely declaring yourselves above it all and labeling both sides equally at fault, to ignore the facts of history and the aggressive impulses of an evil empire, to simply call the arms race a giant misunderstanding and thereby remove yourself from the struggle between right and wrong and good and evil.

The Soviet Union, an evil empire. By the time Reagan delivered this speech I'd joined the staff of Vice President Bush, where, courtesy of the American taxpayer, I found three newspapers a day and about half a dozen magazines a week delivered to my office. I discovered that for months—literally months—I could count on seeing that phrase, "evil empire," referred to in a newspaper or magazine at least once or twice a week. *National Review* and the *Wall Street Journal* applauded it. Nearly every other publication denounced it. But the phrase continued to echo.

Reagan delivered the third speech on June 12, 1987, in Berlin. By then the Soviets found themselves on the defensive, less intent on expanding abroad than on reforming at home. Speaking at the Berlin Wall, the Brandenburg Gate rising behind him, Reagan responded to Mikhail Gorbachev's new policies of *perestroika,* or reform, and *glasnost,* or openness.

We hear much from Moscow about a new policy of reform and openness. Some political prisoners have been released. Certain foreign news broadcasts are no longer being jammed. Some economic enterprises have been permitted to operate with greater freedom from state control. Are these the beginnings of profound changes in the Soviet Union? Or are they token gestures, intended to raise false hopes in the West or to strengthen the Soviet system without changing it . . . ? There is one sign the Soviets can make that would be unmistakable. . . .

General Secretary Gorbachev, if you seek peace, if you seek prosperity for the Soviet Union and Eastern Europe, if you seek liberalization, come here to this gate.

Mr. Gorbachev, open this gate!

Mr. Gorbachev, tear down this wall!

Like the phrase "evil empire," the phrase "tear down this wall" echoed for months.

The last of the four speeches took place on May 31, 1988, during Reagan's visit to Moscow. Standing beneath a gigantic marble bust of Lenin, Reagan addressed several hundred students at Moscow State University.

Freedom is the right to question and change the established way of doing things. It is the continuing revolution of the marketplace. It is the understanding that allows us to recognize shortcomings and seek solutions. It is the right to put forth an idea, scoffed at by the experts, and watch it catch fire among the people. It is the right to follow your dream, or stick to your conscience, even if you're the only one in a sea of doubters.

Freedom is the recognition that no single person, no single authority of government has a monopoly on the truth, but that every individual life is infinitely precious, that every one of us put on this world has been put there for a reason and has something to offer.

The fortieth President, describing freedom to the children of the Soviet apparat. The Cold War was over.

In London, Reagan announced his strategy for dealing with the Soviets; in Orlando, he made the moral case for pursuing it;

in Berlin, he pressed his advantage over the Soviets, who were by then on the defensive; and, in Moscow, he gave a victory speech in which, instead of gloating, he extolled the ideal of liberty.

What strikes me as I reread these four speeches is that each shared the same penetrating, trumpetlike quality as Reagan's 1964 speech for Barry Goldwater. The diplomacy of coexistence and détente, the moral relativism that recognized no real difference between the Soviet Union and the United States, the view that we should manage geopolitical reality rather than transform it—the speeches pierced these old, settled notions the way a trumpet blast pierces din.

This raises a question. You see, although Ronald Reagan wrote that speech on behalf of Barry Goldwater, speechwriters drafted the other four: Tony Dolan wrote the address to the British Parliament and the "evil empire" speech, Josh Gilder the address at Moscow State University, and I myself the speech in the middle, the address in Berlin. How was Reagan able to go right on sounding like Reagan even when other people were composing his words?

For a long time the question puzzled me. I know. It doesn't make sense. I can only tell you it didn't make sense at the time, either. As a member of the speechwriting shop, I saw the entire process from the inside. I knew the procedures for assigning speeches, for pulling together research materials, for editing and circulating drafts, for fact-checking, and for getting final approval from the President. But the whole business struck me as mysterious even so. If you walked from office to office on the speechwriters' hallway, you'd find six people seated at six word processors. But if you looked over their shoulders at their monitors, you'd see that all six were typing words that sounded just like a seventh, Ronald Reagan. It was our job to sound like the President, of course, but that never helped me to understand the way we managed to do so. I felt like an idiot savant. I could perform the trick, but I didn't know how.

I tried for the longest time to figure it out. Then one experience enabled me to see the answer, showing me how, despite a constantly changing stable—although there were only six of us on the staff at any given time, by the end of the administration fourteen people had served as speechwriters—the President always sounded like himself. The experience was writing that speech in the middle, the address in Berlin. Let me describe the experience, then tell you what I learned from it.

—— *The Angry Hausfrau* ——

IN APRIL 1987, when I was assigned to write the speech, the celebrations for the 750th anniversary of the founding of Berlin were already under way. Queen Elizabeth had already visited the city. Mikhail Gorbachev was due in a matter of days. Although the President hadn't been planning to visit Berlin himself, he was going to be in Europe in early June, first visiting Rome, then spending several days in Venice for an economic summit. At the request of the West German government his schedule was adjusted to permit him to stop in Berlin for a few hours on his way back to the United States from Italy. I was told only that the President would be speaking at the Berlin Wall, that he was likely to draw an audience of about ten thousand, and that, given the setting, he probably ought to talk about foreign policy.

In late April, I spent a day and a half in Berlin with the White House advance team, the logistical experts, Secret Service agents, and press officials who went to the site of every presidential visit to make arrangements. All that I had to do in Berlin was find material. When I met the ranking American diplomat in Berlin, I assumed he would give me some.

A stocky man with thick glasses, the diplomat projected an anxious, distracted air throughout our conversation, as if the very prospect of a visit from Ronald Reagan made him nervous. The diplomat gave me quite specific instructions. Almost all of it was in the negative. He was full of ideas about what the President *shouldn't* say. The most left-leaning of all West Germans, the diplomat informed me, West Berliners were intellectually and politically sophisticated. The President would therefore have to watch himself. No chest-thumping. No Soviet-bashing. And no inflammatory statements about the Berlin Wall. West Berliners, the diplomat explained, had long ago gotten used to the structure that encircled them.

The diplomat offered only a couple of positive suggestions. Reagan should mention American efforts to obtain more air routes into West Berlin. And he should play up American support for a plan to turn West Berlin into an international conference center.

After I left the diplomat, several members of the advance team and I were given a flight over the city in a U. S. Air Force helicopter. Although all that remains of the wall these days is paving stones that show where it stood, in 1987 the structure dominated Berlin. From the air, the wall seemed less to cut one city in two than to separate two different modes of existence. On one side lay movement, color, modern architecture, crowded sidewalks, traffic. On the other lay a kind of void. Buildings still exhibited pockmarks from shelling during the war. Cars appeared few and decrepit, pedestrians badly dressed. When we hovered over Spandau Prison, the rambling brick structure in which

Rudolf Hess was still being detained, soldiers at East German guard posts peered up at us through binoculars, rifles over their shoulders. The wall itself, which from West Berlin had seemed a simple concrete structure, was revealed from the air as an intricate complex, the East Berlin side lined with guard posts, dog runs, and row upon row of barbed wire. The pilot drew our attention to pits of raked gravel. If an East German guard ever let anybody slip past him to escape to West Berlin, the pilot told us, the guard would find himself forced to explain the footprints to his commanding officer.

That evening, I broke away from the advance team to join a dozen Berliners for dinner. Our hosts were Dieter and Ingeborg Elz. Germans themselves, the Elzes had retired to Berlin after Dieter completed his career at the World Bank in Washington. Although we had never met, we had friends in common, and the Elzes had offered to put on this dinner party to give me a feel for their city. They had invited Berliners of different walks of life and political outlooks—businessmen, academics, students, homemakers.

We chatted for a while about the weather, German wine, and the cost of housing in Berlin. Then I related what the diplomat told me, explaining that after my flight over the city that afternoon I found it difficult to believe. "Is it true?" I asked. "Have you gotten used to the wall?"

The Elzes and their guests glanced at each other uneasily. I thought I had proven myself just the sort of brash, tactless American the diplomat was afraid the President might seem. Then one man raised an arm and pointed. "My sister lives twenty miles in that direction," he said. "I haven't seen her in more than two decades. Do you think I can get used to that?" Another man spoke. Each morning on his way to work, he explained, he walked past a guard tower. Each morning, the same soldier gazed down at him through binoculars. "That soldier and I speak the same language. We share the same history. But one of us is a zookeeper and the other is an animal, and I am never certain which is which."

Our hostess broke in. A gracious woman, she had suddenly grown angry. Her face was red. She made a fist with one hand and pounded it into the palm of the other. "If this man Gorbachev is serious with his talk of *glasnost* and *perestroika,*" she said, "he can prove it. He can get rid of this wall."

BACK AT the White House I told Tony Dolan, then director of presidential speechwriting, that I intended to adapt Ingeborg Elz's comment, making a call to tear down the Berlin Wall the central passage in the speech. Tony took me across the street from the Old Executive Office Building to the West Wing to sell the idea to the director of communications, Tom Griscom. "The two of you thought you'd have to work real hard to keep me from saying no," Griscom now says. "But when you told me about the trip, particularly this point of learning from some Germans just how much they hated the wall, I thought to myself, 'You know, calling for the wall to be torn down—it might just work.'"

When I sat down to write, I'd like to be able to say, I found myself so inspired that the words simply came to me. It didn't happen that way. *Mr. Gorbachev, tear down this wall.* I couldn't even get that right. In one draft I wrote, "Herr Gorbachev, bring down this wall," using "Herr" because I somehow thought that would please the President's German audience and "bring" because it was the only verb that came to mind. In the next draft I swapped "bring" for "take," writing, "Herr Gorbachev, take down this wall," as if that were some sort of improvement. By the end of the week I'd produced nothing but a first draft even I considered banal. I can still hear the *clomp-clomp-clomp* of Tony Dolan's cowboy boots as he walked down the hallway from his office to mine to toss that draft onto my desk.

"It's no good," Tony said.

"What's wrong with it?" I replied.

"I just told you. It's no good."

"What's no good? Which paragraphs do I need to rewrite? Be a little more specific, Tony."

"The whole thing is no good."

"The whole thing?"

Tony just looked at me.

The following week I produced an acceptable draft. It needed work—the section on arms reductions, for instance, still had to be fleshed out—but it set out the main elements of the address, including the challenge to tear down the wall. On Friday, May 15, the speeches for the President's trip to Rome, Venice, and Berlin, including my draft, were forwarded to the President, and on Monday, May 18, the speechwriters joined him in the Oval Office. My speech was the last we discussed. Tom Griscom asked the President for his comments on my draft. The President replied simply that he liked it.

Now, you might suppose that after hearing the President say he liked his draft a speechwriter would feel so delighted he'd leave it at that. Somehow, it didn't work that way. As a speechwriter you spent your working life watching Reagan, talking about Reagan, reading about Reagan, attempting to inhabit the very mind of Reagan. When you joined him in the Oval Office, you didn't want to hear him say simply that he liked your work. You wanted to get him talking, revealing himself. So you'd go into each meeting with a question or two you hoped would intrigue him. In this meeting, for example, Josh Gilder, who had drafted remarks for the President to deliver at the Vatican, had asked Reagan what role he believed religion might play in the reform of Eastern Europe. The President had responded with a beautiful little disquisition on the need for religious renewal in the Soviet Union itself, exposing an aspect of his thinking none of us had seen. I was hoping for something like that.

"Mr. President," I said, "I learned on the advance trip that your speech will be heard not only in West Berlin but throughout East Germany." Depending on weather conditions, I explained, radios would be able to pick up the speech as far east as Moscow itself. "Is there anything you'd like to say to people on the *other* side of the Berlin Wall?"

The President cocked his head and thought. "Well," he replied, "there's that passage about tearing down the wall. That wall has to come down. That's what I'd like to say to them."

As we speechwriters filed out of the Oval Office, I felt let down. Josh had gotten the President to give him a good two pages of marvelous new material. All I'd gotten the President to do was acknowledge a passage I'd already completed.

I mention this to show what a fool I could be.

I spent a couple of days obsessing over the speech, convinced that each of my feverish rewrites represented an improvement. I suppose I should admit that at one point I actually took "Mr. Gorbachev, tear down this wall" *out*, replacing it with the challenge, in German, to open the Brandenburg Gate, *"Herr Gorbachev, machen Sie dieses Tor auf."*

"What did you do *that* for?" Tony asked.

"You mean you don't get it?" I replied. "Since the audience will be German, the President should deliver his big line in German. And the big line should be about the Brandenburg Gate, not the Berlin Wall, because to the Germans the gate is an even more important symbol than the wall.

Shaking his head, Tony put "Mr. Gorbachev, tear down this wall" right back in.

—— *Squelchfest* ——

WITH THREE WEEKS to go before it was delivered, the speech was circulated to the State Department and the National Security Council. Both attempted to squelch it. The assistant secretary of state for Eastern European affairs challenged the speech by telephone. A senior member of the National Security Council staff protested the speech in memoranda. The ranking American diplomat in Berlin objected to the speech by cable. The draft was naïve. It would raise false hopes. It was clumsy. It was needlessly provocative. State and the NSC submitted their own alternate drafts—my journal records that there were no fewer than seven—including one written by the diplomat in Berlin. In each, the call to tear down the wall was missing.

Now in principle, State and the NSC had no objection to a call for the destruction of the wall. The draft the diplomat in Berlin submitted, for example, contained the line, "One day, this ugly wall will disappear." If the diplomat's line was acceptable, I wondered at first, what was wrong with mine? Then I looked at the diplomat's line once again. "One day?" One day the lion would lie down with the lamb, too, but you wouldn't want to hold your breath. "This ugly wall will disappear?" What did *that* mean?

That the wall would just get up and slink off of its own accord? The wall would disappear only when the Soviets knocked it down or let somebody else knock it down for them, but "this ugly wall will disappear" ignored the question of human agency altogether. What State and the NSC were saying, in effect, was that the President could go right ahead and issue a call for the destruction of the wall—but only if he employed language so vague and euphemistic that everybody could see right away he didn't mean it.

The week the President left for Europe, Tom Griscom began summoning me to his office each time State or the NSC submitted a new objection. Each time, Griscom had me tell him why I believed State and the NSC were wrong and the speech, as I'd written it, was right. When I reached Griscom's office on one occasion, I found Colin Powell, then deputy national security adviser, waiting for me. I was a thirty-year-old who had never held a full-time job outside speechwriting. Powell was a decorated general. After listening to Powell recite all the arguments against the speech in his accustomed forceful manner, however, I heard myself reciting all the arguments in favor of the speech in an equally forceful manner. I could scarcely believe my own tone of voice. Powell looked a little taken aback himself.

A few days before the President was to leave for Europe, Tom Griscom received a call from the chief of staff, Howard Baker, asking Griscom to step down the hall to his office. "I walked in and it was Senator Baker [Baker had served in the Senate before becoming chief of staff] and the secretary of state—just the two of them." Secretary of State George Shultz now objected to the speech. "He said, 'I really think that line about tearing down the wall is going to be an affront to Mr. Gorbachev,'" Griscom recalls. "I told him the speech would put a marker out there. 'Mr. Secretary,' I said, 'The President has commented on this particu-

102

lar line and he's comfortable with it. And I can promise you that
this line will reverberate.' The secretary of state clearly was not
happy, but he accepted it. I think that closed the subject."

It didn't.

When the traveling party reached Italy (I remained in
Washington), the secretary of state objected to the speech once
again, this time to deputy chief of staff Kenneth Duberstein.
"Shultz thought the line was too tough on Gorbachev,"
Duberstein says. On June 5, Duberstein sat the President down
in the garden of the estate in which he was staying, briefed him
on the objections to the speech, then handed him a copy of the
speech, asking him to reread the central passage.

Reagan asked Duberstein's advice. Duberstein replied that he
thought the line about tearing down the wall sounded good. "But
I told him, 'You're President, so you get to decide.' And then,"
Duberstein recalls, "he got that wonderful, knowing smile on his
face, and he said, 'Let's leave it in.'"

The day the President arrived in Berlin, State and the NSC
submitted yet another alternate draft. "They were still at it on the
very morning of the speech," says Tony Dolan. "I'll never forget
it." Yet in the limousine on the way to the Berlin Wall, the
President told Duberstein he was determined to deliver the controversial line. Reagan smiled. "The boys at State are going to kill
me," he said, "but it's the right thing to do."

—— *That Sound* ——

A SPEECHWRITER stumbling around in Berlin, then returning to Washington to flail at his keyboard. A fight over
the draft he produced that raged for three weeks, producing

dozens of memoranda, scores of telephone calls, and meeting after meeting among officials of the State Department, the National Security Council, the White House senior staff, and the speechwriting office. And at the end of this protracted, messy, ugly, irrational process—what? Ronald Reagan, standing before the Berlin Wall, sounding like—well, sounding like Ronald Reagan.

How was that possible?

The sound of the trumpet. That, I decided, was the explanation.

Until he became President, Ronald Reagan wrote most of his own speeches, developing his own distinctive voice or sound. Simplicity, directness, the diction of ordinary American speech, a certain sense of energy—all contributed to the Reagan sound. But the dominant element, the feature that gave Reagan's speeches their trumpetlike quality, was his insistence on telling the truth. We might wish the Soviet Union had changed over the years, becoming a regime of simple bureaucrats with whom the West could comfortably coexist. It hadn't. Instead it had incarcerated political and religious prisoners in a vast internal gulag, armed and funded subversive organizations around the world, and supported police states throughout Eastern Europe, building a wall around West Berlin not to keep West Berliners out of East Germany but to prevent East Germans from escaping to West Berlin. The Soviet Union had remained an evil empire—and Reagan insisted on saying just that. No more euphemisms. No more wishful thinking. The truth.

Now, I couldn't have put it that way for you when I was assigned to write the Berlin Wall speech—I didn't think any of this through in quite these terms until afterward. But I didn't need an acute analytical grasp of Reagan's speaking style. None of us speechwriters did. We just had to be able to recognize it. When

I had dinner with the Elzes, I knew the moment Ingeborg Elz made her remark about getting rid of the wall that Reagan would have responded to it just as I did, admiring the passion and decency it conveyed. Get rid of the wall? That sounded like him. And although I may have flailed around at my word processor back in Washington when I tried to adapt the remark, building a speech around it, I felt sure I'd be able to produce a compelling draft sooner or later. All I had to do was match the trumpetlike sound that Reagan himself always made.

And that is my point. Ronald Reagan was able to go right on sounding like Ronald Reagan after becoming President because he'd spent so many years sounding like Ronald Reagan before becoming President. We speechwriters were never attempting to fabricate an image. We were only attempting to produce work that met the standard Reagan himself had long ago established. And when the State Department and the National Security Council tried to block my draft by submitting alternate drafts, they weakened their own case. Their drafts were drab. They were bureaucratic. They lacked conviction. They failed to match the trumpetlike sound of Ronald Reagan.

—— *Berlin Again* ——

THIS MAY SOUND like another odd admission coming from a speechwriter, but in certain moods I found myself wondering whether even Reagan's big speeches really mattered. Unlike, say, my remarks for the Ding Darling Duck Stamp, speeches such as the four I've discussed here—the address to the British Parliament, the "evil empire" speech, the Berlin Wall address, and the address at Moscow State University—all received extensive coverage in the media. For us speechwriters, that was almost reward enough in itself. After one of us scored—if a speech you'd written was excerpted in the *New York Times* "Quotation of the Day" you counted yourself especially successful—there would be high fives all around. Reagan's big speeches had an obvious impact on American politics, of course, helping the President retain widespread support. But did they have any effect outside the United States?

From time to time, we speechwriters would hear that the President's speeches annoyed the Soviet leadership. Word reached us, for example, that, in several of his meetings with the secretary of state, Soviet ambassador Anatoly Dobrynin had railed against Reagan's speeches. And when Anatoly Shcharansky visited the White House in 1986, we learned that excerpts from the President's speeches had found their way into the Soviet gulag. A Jewish refusenik who had spent eight years in Soviet prisons, Shcharansky told the President that prisoners passed from hand to hand tiny slips of paper onto which Reagan's words had been copied in minute script. Dobrynin red in the face; prisoners passing slips of paper back and forth. We cherished these stories, of course, but they hardly suggested the President's speeches were causing the Soviet empire to crack.

After the Berlin Wall address, only a single piece of evidence that the speech had produced any practical results ever came to my attention. At lunch in the White House mess a week after the speech, a member of the National Security Council staff told me that our intelligence services had picked up unusual cable traffic between Moscow and East Germany. The Soviets, the cable traffic showed, wanted the East Germans to make the Berlin Wall less offensive to the West, opening more checkpoints or easing travel restrictions on people who wanted to visit their relatives. "Each generation of Soviet leaders," said the NSC staffer, who, after opposing the Berlin Wall address, had changed his mind about it, "needs to be reminded that the wall is a public relations disaster." I'd have called the wall an affront to human decency, not just a public relations disaster, but I understood what he meant.

Yet that was all I'd ever heard. A suggestion that the speech might have prompted some cable traffic. After that, nothing.

Then, last year, I returned to Berlin.

I was helping Fox News produce a documentary marking the fifteenth anniversary of the Berlin Wall address. As the cameras rolled I told Tony Snow, the Fox News anchorman who narrated the documentary, about the fight over the speech, then joined Snow in visiting the spot in front of the Brandenburg Gate where Reagan had spoken. After the day's shooting, Snow and I walked through the Brandenburg Gate to the former East Berlin, a place that once appeared drab and lifeless but now pulsed with color and energy. (In a building that had once housed a branch of the old Communist government, Snow and I discovered a particularly vivid demonstration of the victory of free markets: a Rolls-Royce dealership.) Yet the most memorable part of the trip took place at the home of Dieter and Ingeborg Elz. Re-creating their 1987 dinner party for me, the Elzes had brought together as many of their original guests as they could, then added two new guests: Ulrike Marschinke, a woman who helps Dieter with a journal he edits, and Otto Bammel, a retired West German diplomat. Fifteen years before, Marschinke and Bammel had been living in East Berlin.

"When I heard Mr. Reagan say, 'Mr. Gorbachev, tear down this wall,' " Marschinke explained for the television cameras, "I thought to myself, 'What a strange idea!' I only knew the world with the wall. In the East, the Communist Party; in the West, the rest of the world. I couldn't imagine how it would work to live without the wall. It was impossible for me to understand what would happen."

After the taping, Otto Bammel, the retired diplomat, took me aside to tell me what he had witnessed in November 1989. Representing the government of West Germany, Bammel was living with his wife and two sons, both of whom were in their early twenties, in an East Berlin home just a few hundred yards from the wall. During the evening of November 9, as the East German state council met in emergency session—a few days earlier there had been peaceful but massive demonstrations throughout East Berlin—Bammel and his oldest son, Karsten, watched television as an East German official held a press conference.

"It was so boring," Bammel said, "that I finally couldn't take any more. So I said, 'Karsten, you listen to the rest. I'm going into the kitchen for something to eat.' Ten minutes later Karsten came to me and said, 'The official just announced everyone can go through the wall! It's a decision made by the state council!' I didn't believe this could happen. It was an unbelievable event." Certain that his son had somehow misunderstood, Bammel took his wife to the home of a neighbor, where they were expected for dinner.

"When we got back at midnight we saw that our boys were still out," Bammel continued. "And we were surprised that there were so many cars driving within the city, but where the traffic goes and why it was, we did not know. We went to bed. When we got up at seven o'clock the next morning, our boys were sleeping in their beds. We saw a piece of paper on our kitchen table from our youngest boy, Jens, telling us, 'I crossed the wall. I jumped over the wall at the Brandenburg Gate with my friends. I took my East Berlin friends with me.'

"I said to my wife, 'Something is wrong.' Without eating we took our bicycles and went to the border. And that was the first time we saw what happened in the night. There were people crossing the border on foot and in cars and on bicycles and motorbikes. It was just overwhelming. Nobody expected it. Nobody had the idea that it could happen. The joy about this event was just overwhelming all other thoughts. This was so joyful and so unbelievable."

As I looked over my notes in my hotel room that evening, I was struck by the way Marschinke and Bammel had both used language suggesting a sense of incredulity or unreality. The wall, both had implied, had seemed so real, solid, and immovable—such a fixed part of everyday life, of the East German state, and of the entire Communist outlook and philosophy—that the very idea of tearing it down had by contrast seemed strange and fantastic. Even when the wall had ceased to function, Bammel, a professional diplomat, had had trouble believing it until he had seen East Germans freely crossing the border with his own eyes.

Ronald Reagan, I recognized in that Berlin hotel room, had given something to people in the East, something difficult to describe but tangible all the same. In predicting that Communism would end up on the ash-heap of history, in describing the Soviet Union as an evil empire—in insisting that the West remained fundamentally vibrant and good, the Soviet Union backward and corrupt—Reagan had spoken the unspeakable. He had done what no one could do. And he had thus created for people in the East a new space for thought and feeling, a new sense of the possible. If an American President could call on the leader of the Soviet

Union to tear down the Berlin Wall—if that could happen, if it were true—then what else might prove possible?

Reagan had never been alone in calling for freedom. Pope John Paul II, Lech Walesa, Vaclav Havel, and others had all denounced the Communist regimes of Eastern Europe, demanding human rights. Yet Reagan's voice had always proven among the most compelling and insistent. "That wall has to come down," he'd replied when I asked what message he wanted to convey to people in the East. "That's what I'd like to say to them."

Did Reagan's speeches matter? Enough to change the world.

—— *Sticking Up* ——

WORDS MATTER. Stick up for your beliefs, speak with conviction, and remember that people respond to the truth—even those who expressed doubts about the Berlin Wall speech saw as soon as the President delivered it that he'd been correct to do so. As Tom Griscom recalls, "The secretary of state found me after the speech and said, 'You were right.' " Thinking over this lesson in the White House, I'll admit, I couldn't see what good it would ever do me. When Ronald Reagan spoke, millions listened. When I spoke, even waiters paid no attention. (To this day I have trouble getting a second cup of coffee.) As it turns out, though, I employ the lesson all the time.

To show you how, let me describe a brief exchange that took place on the PBS television program I host, *Uncommon Knowledge*. Discussing the end of the Cold War, the journalist Christopher Hitchens argued that Eastern European intellectuals such as Vaclav Havel brought about the Velvet Revolution of 1989 entirely on their own, as if Ronald Reagan had had nothing to do with it. I challenged Hitchens.

> HITCHENS: Vaclav Havel . . . mentions as his cultural icons . . . Frank Zappa and the Velvet Underground. . . . [And] in an interview with me once said, "I consider myself a man of the Sixties and of [19]68." There were innumerable symptoms in the 1989 revolution of a sort of delayed element of the '68 one.
>
> ME: But you're not going to assert that the Velvet Revolution in Czechoslovakia . . . could have taken place without Ronald Reagan beefing up American military might.
>
> HITCHENS: I would certainly assert that. I think that—
>
> ME: You *would* assert that?
>
> HITCHENS: I think that can be asserted pretty confidently.
>
> ME: You're saying that the Soviet Union would have fallen because Vaclav Havel liked Frank Zappa?
>
> HITCHENS: No. I'm saying that the movements of democracy in Eastern Europe were indigenous, yes, and would have prevailed.

When you're hosting a television program, at some point you have to give up on one line of inquiry to move on to the next, and that is what I did now. First, though, I gave Hitchens a look. Reagan had nothing to do with the events of 1989? Right. Hitchens, for his part, appeared taken aback, as if he'd never before heard anyone stick up for Ronald Reagan.

Why mention this exchange? Only because it nearly didn't

take place. Christopher Hitchens is supremely articulate. In defending Reagan, I ran the risk that he'd reply with a witty riposte, leaving me spluttering for three or four seconds, a long time in television, especially when you're the one who's red in the face. For an instant I considered keeping my mouth shut, permitting Hitchens's assertion to pass. Then I went right ahead and stuck up for Reagan anyway. The moment I did, I felt proud of myself.

Which is my point. When I worked at the White House, I valued words for the effect they had on those who heard them. Reagan would give a speech on economics. Tax rates would go down. He'd give a speech on defense. Spending on the military would go up. But since then I've also come to value words for the effect they have on those who speak them. When, if only in a small way, I stick up for my beliefs, as in my exchange with Christopher Hitchens, my words do something to me even if nobody else pays any attention. They firm up my self-respect. They make me a better person.

Once I'd grasped this, I saw that Reagan himself must have understood it all along. He'd spent years speaking out on issues even when he'd had no reason to believe anything he said would ever make a difference. During the Johnson administration, for example, Reagan had opposed big government in speech after speech even as Johnson had enacted the Great Society, the most massive expansion of the federal government since the New Deal. And even once he'd become President, Reagan had never quite been able to tell what effect his words might have. He'd admitted as much a few years after leaving office when he remarked that he'd never expected the Berlin Wall to come down as soon as it had. Even Reagan—even the Great Communicator—seems to have felt just as unsure as I did about the impact his speeches might have outside the United States. As he'd overruled the State

Department and the National Security Council to deliver the Berlin Wall address, all Reagan had known for certain was what he'd told Kenneth Duberstein in the limousine on the way to the wall itself: "It's the right thing to do."

Every now and again, I'll learn that something I've said has had an effect—after that exchange with Christopher Hitchens, a couple of viewers sent me e-mails to say they'd appreciated my defense of the Gipper. But I've found that even when I have no idea whether anyone is ever listening, which, of course, is most of the time, sticking up for my beliefs is its own reward.

----------- *Five* -----------

AT THE BIG DESK
IN THE
MASTER BEDROOM

You Have a Head. Use It.

JOURNAL ENTRY, MAY 2002:

*"My most vivid memory of Dad?" Michael Reagan asked when we spoke today. "Easy. Back before he became governor, he did a lot of his work at home—that would be in the house in Pacific Palisades. When I'd get back from school in the afternoon, I'd toss down my books and go into the master bedroom to say hello. Dad had a big desk in there, and he was always at that desk, writing. Not almost *always*. Always."*

—— *Expert Opinion* ——

JUST HOW qualified do you have to be to form your own opinion on the issues, cast your vote responsibly, or run for office? Anybody can mouth off on tax rates or foreign affairs, of course.

But should the rest of us listen when he does? When I landed at the White House, I found these questions on my mind a lot. They'd been bothering me since college, when the President of Dartmouth, John George Kemeny, had gotten me thinking about them.

By any definition, Kemeny counted as a genius. An immigrant from Hungary, he had so distinguished himself in mathematics that while still a Princeton undergraduate he had been recruited to work on the Manhattan Project. Returning to Princeton, he had become an assistant to Albert Einstein, earning a doctorate at twenty-three. Kemeny had gone on to make major contributions in mathematics, philosophy, and sociology, and, fascinated by computers, to invent BASIC, one of the first computer languages.

Witty and charming, Kemeny had a dapper little mustache, used a cigarette holder, and spoke with an accent that gave him an air of European urbanity. Learning that he reserved an hour each week when anyone in the college could drop in on him, my freshman-year roommate and I managed to walk in the door of Parkhurst Hall, the administration building, three or four times, but we regarded Kemeny with such awe that it took us a couple of months to work up the courage to mount the stairs to the President's office. When his secretary showed us in, Kemeny smiled affably, rising from his desk, then directed us to armchairs near the fireplace. Only once we were seated did my roommate and I realize we had nothing to say. Blushing, we blurted out the truth, admitting that we'd stopped by just to be able to say we'd met him. Kemeny chuckled. Then he treated us with perfect courtesy, asking where each of us had grown up, which courses we were taking, and whether we had any suggestions for improving the administration of the college. We could hardly believe it.

Throughout the conversation, John George Kemeny sustained the fiction that we were his equals.

Kemeny made me proud of attending the institution over which he presided. He also made me uneasy. You see, he would deliver a speech at every major college function—convocation at the beginning of the academic year, commencement at the end, and two or three other big events in between—and in nearly every speech he would discuss the increasing complexity of American life. At the beginning of the twentieth century, the argument went, the United States remained largely agrarian. You planted your crops and milked your cows. Since that was about all there was to it, most people could manage on their own. In the urban, industrial society the United States had since become, by contrast, life had grown intricate and interdependent. The economy now had to be managed, the environment protected, and a vast complex of laws and regulations constantly adjusted and revised. Our duty as students, Kemeny advised us, was to become expert enough in our chosen fields to lend a hand.

I knew without his spelling it out what kinds of experts Kemeny had in mind. Physicists, biologists, computer scientists, engineers, sociologists, economists, lawyers—they'd all know how to break propositions into their component parts, conduct studies, run data through computers, and come up with specific solutions to the nation's problems. No matter how complex life in America became, they'd always prove of use. Me? I was an English major. I liked Chaucer and Shakespeare, poets who, as of my freshman year, had been dead, respectively, for 575 and 359 years. Of what use would I ever prove?

Soon I began to worry not just about my own place in an increasingly complex America, but about the place of democracy itself. Throughout our history, I learned in a couple of courses in political science, many had considered government too complicated for ordinary people. During the colonial period, for example, the Tories had insisted that government remain the province of royal governors, landed gentlemen, and rich merchants, men of refinement and means. The common people? Ignorant rabble. Incapable of thinking outside the restrictive categories of class, the Tories, I saw, had been snobs, so I'd found their position easy enough to dismiss. Yet during the twentieth century the old anti-democratic argument had acquired a new relevance and credibility. Now government really *was* too complicated for ordinary people. John George Kemeny suggested as much himself.

When I left Dartmouth for Oxford, my education in the increasing complexity of American life continued. Studying economics, I found that my tutors all subscribed to one version or another of the Keynesian model. In any modern industrial society like that of the United States, they taught me, the free market often failed, forcing government to intervene, typically by increasing public spending. Yet such interventions proved tricky. If it intervened too late, government might prolong a recession; if too early, it might create inflation. Who would decide? Who would constantly monitor the economy, and then, at just the right moment, tell the politicians to increase public spending? The experts, of course. At Dartmouth, at Oxford, everywhere I turned—experts.

By the time I reached the White House, I'd begun to suppose that the ideal of democracy was being gradually hollowed out. Since Americans had spent two centuries venerating democracy, nobody would come right out and say so. But the belief that the insights of ordinary Americans were just as valid as those of anybody else had become anachronistic. To participate in public life—really participate, helping to establish the nation's

117

agenda—you had to be an economist, sociologist, or some other species of expert.

This brings me to the fortieth chief executive. Ronald Reagan proved me mistaken.

—— *Some Dunce* ——

WHEN I WENT to work for him, I knew so little about Reagan's background that I assumed he must have been a conservative Republican at least as long as he'd been active in politics—since he'd delivered his famous speech on behalf of Barry Goldwater when I was in first grade, he'd certainly been a conservative Republican as long as *I* could remember. Although I knew Reagan had once thought of himself as a Democrat, admiring FDR, I figured that had been an adolescent infatuation. A lifelong Republican myself, I took it for granted that Reagan had become a Republican at about the same time he had become an adult.

Then one afternoon as I was flipping through old materials in the research office, I came across newspaper clippings that described speeches Reagan had delivered on behalf of Harry Truman during the presidential campaign of 1948, when Reagan was already in his late thirties. Heading an organization called "Hollywood for Truman," Reagan had described himself as a Democrat—a *liberal* Democrat, mind you. That made me stop and scratch my head. A liberal Democrat in 1948, a conservative Republican in 1964. If Reagan hadn't simply grown out of the Democratic Party, how *had* he become a Republican?

Talking to people who had known him during those early years, including his children Maureen and Michael, I learned the answer. Reagan had traveled from left to right by thinking.

When he wasn't in a film or television studio, on the road, or at his ranch, caring for his horses, everyone with whom I spoke agreed, Reagan was studying a book, leafing through a political journal, or laboring over index cards or a legal pad, pen in hand.

A reader all his life—"He wouldn't feel right without having something to read," Ron Reagan now says, "so he'd always have a book going"—Reagan gave himself a political education by poring over works of history, economics, and biography. A couple of Reagan hands recalled seeing him reading Milton Friedman's classic, *Capitalism and Freedom*, as he was preparing to run for governor in 1966, pulling the book out of his jacket pocket between meetings. Others had heard him discuss *Witness*, the autobiography of Whittaker Chambers. In one passage, Chambers describes finding himself struck by the beauty of his infant daughter's ears.

> My daughter was in her high chair. I was watching her eat. She was the most miraculous thing that had ever happened in my life . . . My eye came to rest on the delicate convolutions of her ear—those intricate, perfect ears. The thought passed through my mind: "No, those ears were not created by any chance coming together of atoms in nature (the Communist view). They could have been created only by immense design . . ." I did not then know that, at that moment, the finger of God was first laid upon my forehead.

Reagan could recite the passage almost word for word. An avid reader not only of books but of newspapers and magazines, Reagan always pored over the conservative journals *Human Events*, *American Spectator*, and *National Review* from cover to cover. When I asked William F. Buckley Jr., the founder and editor of *National*

Review, just what he and his friend the President talked about when they saw each other or spoke over the telephone, Buckley surprised me, replying that the two most prominent conservatives in the country seldom discussed public policy. "He reads *National Review* so thoroughly," Buckley said only half-jokingly, "that there's never anything I can tell him."

Although widely derided as an "amiable dunce," to use Clark Clifford's phrase, Reagan was content to let his reading habits go unremarked. Over lunch one afternoon in the White House mess, Marlin Fitzwater, then the press secretary, described a telling incident. Traveling with the President to Europe, Fitzwater noticed that Reagan had brought along a stack of books. "As I recall," Fitzwater now says, "most were biographies of American leaders, and a couple were fictional best-sellers at the time." Fitzwater asked the President if he could jot down a list of the books to hand out to the press. "He said it didn't really matter to him, but it was probably unnecessary." Able to take a presidential hint, Fitzwater let the matter drop.

Reagan wrote as steadily as he read. Before becoming governor, I learned, Reagan had composed his material himself, writing all the speeches he had delivered as a political activist, as President of the Screen Actors Guild, and as the celebrity spokesman for General Electric. He'd had help on his autobiography, *Where's the Rest of Me?*, but chiefly in editing the manuscript he had already produced. After becoming governor, Reagan had continued to compose much of his material, writing some speeches himself while often rewriting the speeches his staff prepared for him. "Sometimes he'd only change one word of my stuff," Lyn Nofziger, who wrote speeches for Reagan during this period, says. "But a lot of times he'd change just about everything *except* one word." After stepping down as governor, Reagan had turned over a lot of his writing responsibilities to Peter Hannaford, who, with Michael Deaver, formed a public relations firm to coordinate Reagan's speaking engagements, daily radio spot, and weekly newspaper column. Yet according to Hannaford himself, Reagan had still composed most of his speeches, most of his radio talks, and a substantial portion of his newspaper columns.

One afternoon I sat down with a pencil and paper to take a few stabs at adding up all the material Reagan had composed. My estimates could only prove approximate, of course—nobody knew quite how many speeches Reagan had delivered before becoming governor or how much of the material in each of those speeches had been original, how much repeated. Yet even by my *lowest* estimate, Reagan must have composed about half a million words. As best I could work it out, the "amiable dunce" had done more writing than any chief executive since Woodrow Wilson.

Ronald Reagan, the author of his own material. This was an idea that took some getting used to, I'll admit. If only the President had time, he could write all his speeches himself—that, I learned when I joined the staff, was the correct reply to any reporter who asked about my job. It sounded good, and I certainly said it often enough. But I couldn't help wondering if it were true.

Then in late 1983, James Watt, the secretary of the interior, made a remark belittling affirmative action, and, in the furor that followed, Watt found himself forced to resign. The President felt for the former secretary—Watt may have said something stupid, but he'd been a loyal member of the administration all the same—so Reagan decided to devote one of his weekly radio talks to a review of Watt's accomplishments. Never learning of this decision, the speechwriting office sent the President a radio talk that dealt with the economy, not Watt. The President solved this problem by writing a radio talk of his own.

When the President's draft reached the speechwriting office for fact-checking, we passed it around. Single-spaced on a yellow legal pad, the draft ran to two pages in the President's own handwriting. It was relaxed and conversational, a perfect little piece of oratory. But what struck me was the way it looked. It was clean. The draft contained only two or three rewrites, each an instance in which the President had crossed out only a single word or phrase. Whereas we speechwriters always rewrote extensively, crossing out so many paragraphs, adding and dropping so many words and phrases, and scribbling so many new passages in the margins that our secretaries often had trouble making out our changes, Ronald Reagan had simply sat down and written what he wanted to say.

It was true. If only the President had the time, he *could* write all his speeches himself.

—— *Left to Right* ——

THE STEPS Reagan followed in thinking his way from left to right? I was never able to trace them. Too few of his writings had survived. Whereas some Presidents—John Adams comes to mind—seem to have begun worrying at an early stage in life about the way historians would evaluate them, squirreling away copies of nearly everything they wrote, Ronald Reagan doesn't seem to have cared about the opinions of historians at any stage of his life, young, middle-aged, or old, so he never squirreled away copies of anything he wrote. Although the research office had plenty of his speeches, radio talks, and newspaper columns from Reagan's 1966 election as governor of California onward, materials for the earlier years proved sparse.

When had Reagan first begun inveighing against high taxes? How had anti-Communism become one of his defining causes? No one, I realized, would ever be able to say. Yet certain experiences had impressed themselves on Reagan, as he himself proved by continuing to talk about them in the White House. Facing a 91 percent tax bracket after World War II, for example, Reagan had developed qualms about progressive rates of taxation. As I heard him tell it, he had avoided the confiscatory rate simply by limiting the number of pictures he made each year, a practice he could easily afford. Yet in prompting stars such as Reagan himself to make fewer pictures, he had seen, the tax code punished soundmen, cameramen, set designers, makeup artists, and a lot of other ordinary working people. At about the same time, Reagan had found himself startled and then angered by the behavior of Communists and Soviet sympathizers. Giving speeches as President of the Screen Actors Guild, he had noticed that although when he talked about the importance of SAG everyone applauded, when he praised the United States, some—the fellow travelers—sat on their hands. Describing these scenes more than three decades later, Reagan still sounded disbelieving.

We all have experiences that impress themselves on us, of course. But because Reagan was always reading and writing, I decided, he had been able to reflect on his experiences with unusual depth, placing them in the context of the political, moral, and intellectual struggles of the day. For him, I saw, cutting back on the number of pictures he made in order to avoid confiscatory tax rates had represented more than an exasperating necessity. It had represented an illustration of the harm that could be done by an overweening federal government, a specific instance from which Reagan, prompted by reading, among others, Milton Friedman, had drawn the general conclusion that the expansion of the welfare state posed a threat to free markets. Likewise, I concluded, Reagan's first experience of Communists and fellow travelers. Watching them demonstrate their disdain for the United States had amounted to more than a series of disturbing encounters. It had amounted to a proof, a set of specific instances from which Reagan, informed by reading, among others, Whittaker Chambers, had drawn the general conclusion that Communism would always remain fundamentally opposed to self-government and human rights.

Examining Reagan's old speeches, radio talks, and newspaper columns, I couldn't quite get over it. There was just so much that Reagan had *known*. He'd known a lot about economics, taxation, regulation, the environment, and energy policy. He'd known the size and budgets of both the Soviet and American military forces over the last couple of decades, and he'd known how much farmers had been paid over the same period for a bushel of wheat. He'd known the history of taxation in ancient Rome (as taxes went up, Rome declined), the history of the Republic of China, and the history of every major treaty of the last century (they'd almost all been broken). He'd been familiar with the writings of the Founders, quoting George Washington, Thomas Jefferson, and others again and again (he'd had a special fondness for John Winthrop, the Puritan leader who declared, "We shall be as a city upon a hill"). He'd understood the importance of interpreting the Constitution according to original intent, and he'd been able to cite the cases in which the activism of the Warren Court had proven the most egregious. Every issue of the day, every pertinent fact, every enduring aspect of the American tradition—Reagan, I saw, had passed them all through the medium of his own intellect, engaging in the act that more than any other enables a man to learn his own mind: putting his thoughts on paper. "He knows so little and accomplishes so much," Bud McFarlane, Reagan's national security adviser, once remarked. Silly him. Behind the chief executive's unassuming manner lay an active mind.

—— *Bake-Off* ——

BECAUSE DARTMOUTH was located in New Hampshire, the state with the first presidential primary, sooner or later every presidential candidate gave a speech on the campus. Among those who visited during the election year of 1976, when I was a sophomore, two stood out. One was Jimmy Carter, then months from capturing the White House. Carter stressed the training he'd received in the Navy, presenting himself as an engineer, a dispassionate, objective problem solver—in other words, an expert. I came to think of Carter as the Kemeny candidate. (And it came as no surprise when three years later Carter named Kemeny to head the investigation of the nuclear accident at Three Mile Island.) The other was Ronald Reagan, then engaged in his unsuccessful primary challenge to Gerald Ford. By contrast with Carter, Reagan made no attempt to present himself as especially intelligent or well qualified, instead talking about his stands as if they amounted to no more than common sense. Reagan's remarks, his sense of humor, and his relaxed, genial demeanor, even when a group of left-wing students heckled him—all suggested that Reagan saw himself not as an expert, in possession of superior skills and knowledge, but only as an ordinary American promising to do his best. Carter the expert, Reagan the ordinary American. The men represented such opposites that their candidacies might almost have been designed as a neat public experiment, as simple and straightforward as a television commercial intended to show which of two mixes produced a moister cake.

Once he became chief executive, Carter spent countless hours mastering every aspect of federal policy—following him as I continued my studies at Dartmouth and then Oxford, I found myself marveling that Carter seemed to be pulling even more all-nighters than I. Yet Carter presided over the worst economic disarray since the Great Depression. And he looked on in befuddlement as the Soviets expanded their influence in Central America and invaded Afghanistan, then watched in bewilderment as radical mullahs staged a coup in Iran, taking 52 Americans hostage. Once he succeeded Carter as chief executive, Reagan, by contrast, never attempted to master every aspect of federal policy, taking it easy instead. All-nighters? Reagan? Even while still in England, watching Reagan's first year in office from my dank Oxford cottage, I recognized that the very idea was preposterous. Yet in reviving the economy and placing the Soviets on the defensive, Reagan so outperformed Carter that the contest wasn't even close.

"I would rather be governed by the first 2,000 names in the Boston telephone directory," William F. Buckley Jr., once remarked, "than by the faculty of Harvard." Ronald Reagan would have agreed. In some ways, of course, Reagan was exceptional. How many people have the talent to become movie stars? But in most ways Reagan proved about as ordinary as ordinary gets. He had attended not a prestigious university in the East but a small Midwestern college, by his own account devoting more attention to football and student drama than to his studies. His tastes ran not to Shakespeare or Ibsen but to Hollywood movies, not to grand opera but to popular crooners such as Bing Crosby and

Frank Sinatra. He was just the kind of person you'd expect to find by placing your finger down at random in the telephone book. And if Reagan could draw conclusions about the issues of the day that proved at least as good as those of the experts, then anybody could. "They say the world has become too complex for simple answers," Reagan said in that 1964 speech on behalf of Barry Goldwater. "They are wrong. There are no easy answers, but there are simple answers."

Simple answers. Cut taxes, roll back regulations, restrain the growth of the federal government, rebuild our defenses, stand up to the Soviets. After an education in which I'd learned that the only good answer was a complicated answer, I'm not sure I'd have believed it if I hadn't been working where I could see it with my own eyes. Yet there in the Oval Office sat Ronald Reagan, insisting on simple answers—and changing the world.

—— *Beats Me* ——

THE CHIEF EXECUTIVE'S belief in simple answers, I saw, had a couple of implications for the way I lived my life. The most immediate was that in performing my job as a speechwriter, I could relax.

You see, when I started work at the White House, I'd assumed a speechwriter merely took the ideas that some wise person had already decided a speech should contain, then arranged them, adding a couple of rhetorical flourishes while making certain the material read clearly and grammatically. In short, I'd supposed a speechwriter concerned himself with style but not substance. George H. W. Bush had straightened me out.

In my first speech meeting with him, the Vice President waved me to a chair across from his desk, looked over the list of speeches I'd handed him, then nodded, indicating I could begin. Referring to the first speech on the list, I informed Bush that in a certain hotel ballroom on a certain date he'd be addressing a certain trade association—if memory serves, the National Association of Manufacturers. "What would you like to say to the manufacturers, sir?" I asked, poised to make notes.

"Beats me," the Vice President replied. "What would you suggest?"

What would *I* suggest? That the Vice President would ask the advice of an inexperienced twenty-five-year-old had never entered my mind. I cleared my throat, tugged at the knot in my tie, looked over the list of speeches, studied the empty pages of my notebook, and then, recognizing that I had to say *something,* muttered a few words about the need for manufacturers to support the administration's economic program in spite of the recession. "You might want to urge them to 'stay the course,' " I mumbled, quoting a phrase I'd discovered when I leafed through a few of the President's speeches the day before.

"Stay the course?" the Vice President said. "Yeah, I guess that'll do. What's the speech after that?"

Now, the Vice President wasn't being quite as flip in this encounter as he may sound. Whatever I came up with, he knew, half a dozen members of his staff would look my material over, correcting and improving it before the text reached him. In the meantime, he wanted to make a point. He was a busy man. He had no time to prepare for speeches by performing background reading, talking to representatives of the organizations before which he'd be appearing, or discussing likely topics with his staff.

That, he wanted to make clear, was my job. Instead of serving as some sort of in-house prose stylist, I was to take responsibility for the substance of his speeches.

On the President's speechwriting staff, I'd learned when I left Bush's staff to become a presidential speechwriter, the case was much the same. Although the guidance we speechwriters received would vary from one speech to another, often it would amount to no more than the date and location of the speech, the name of the organization the President would be addressing, and a one- or two-word description of the theme—"economics," "tax cuts," or "foreign policy." The rest? That would be up to us. The senior staff would make changes on each speech as it circulated, of course, and then the President himself would mark up the final text. But the speech's fundamental message, the arguments it employed, the statistics and anecdotes it presented—all would be the responsibility of a member of the speechwriting staff.

I had a couple of responses to this. The first was that I liked it. I liked it a lot. All of us speechwriters did. The latitude we enjoyed enabled us to have a lot of fun. After learning one afternoon that there was an effort afoot in Congress to raise taxes, for instance, Josh Gilder sat down at his word processor and then, a wicked gleam in his eye, composed an insert for a speech he'd written. "My veto pen is drawn and ready," the President said when he delivered the speech the next day, "and I have only one thing to say to the tax increasers." Reagan paused for a full, rounded beat, his eyes alight with pleasure. "Go ahead. Make my day." A couple of hours later, the effort to raise taxes collapsed. Josh and I exchanged high fives.

In a second instance, I found myself assigned to write a speech on education policy. Although officials at the Department of

Education could tell me about this or that individual program, I discovered, they were completely incapable of describing a coherent, overall framework or approach.

"You're telling me the administration doesn't *have* an education policy?" Dick Darman, then the director of communications, asked when I telephoned to discuss the problem.

"That's about what it comes down to."

"Then make one up," Darman said, and hung up.

I placed a couple of dozen telephone calls, locating several people in the administration who knew a lot about education, including William Bennett, who would later become secretary of education. Then I huddled with them for a few days, devising the education policy I figured the President would have come up with if he'd been in my place. When the President delivered the speech, Josh and I exchanged high fives all over again.

Yet my second response to the latitude we speechwriters exercised proved just the reverse of the first. It made me queasy. When I arrived in Washington, I thought I'd be lucky to land a job writing speeches for the postmaster general. Now here I was framing policy for the chief executive. Who did I think I was? When I read someplace that a lot of people found themselves haunted by the irrational thought that they were frauds, I recognized myself immediately—with the difference, of course, that I really was a fraud. One day a White House security officer called on me. He only wanted to know why I'd begun sending out letters on the apple-green stationery that was reserved for the use of the President, and we quickly established a) that until that moment I'd never known apple-green stationery was different from any other, and b) that somebody in the downstairs supply room had handed my secretary a stack of the stuff by mistake. But until the security officer left my office, taking my apple-green stationery with him, I kept thinking he'd drop the stationery ruse to question me instead about impersonating a speechwriter.

JOURNAL ENTRY, MARCH 1985:

Found this passage in an essay by Lionel Trilling. Trilling was writing about George Orwell, but he might as well have had in mind that other great anti-Communist, Ronald Reagan.

If we ask what it is he stands for, what he is the figure of, the answer is: the virtue of not being a genius, of fronting the world with nothing more than one's simple, direct, undeceived intelligence, and a respect for the powers one does have, and the work one undertakes to do. . . . He is not a genius—what a relief! What an encouragement. For he communicates to us the sense that what he has done, any one of us could do.

At last I saw that what I did as a speechwriter was no different from what the President himself had always done. Without any special qualifications, Reagan had spent several decades reading up on the issues of the day, then writing about them. The chief executive had a good head, I saw, but not a trained mind. The same went for me. If he could write speeches, well, then, I figured, so could I. Ronald Reagan made me feel like an honest man at last.

Eventually I even learned to see my lack of qualifications as a qualification in itself. Whenever an expert confused me—when, for example, an official from the Treasury Department began using technical terms in briefing me for a budget speech—I'd stop him. I was only a layman, I'd explain. But then so were the President and nearly all those who would be in his audience. And if I was unable to understand what the expert was saying, then the expert could feel pretty certain the President and his audience would be unable to understand it themselves. The expert would look at me, scowl, and then do just what I wanted, restating his argument in plain language. Technical gobbledygook, professional jargon, expert rigmarole—my ignorance kept them out of my work.

This brings me back to the question with which I started this chapter, the place of democracy in the United States. If the most immediate implication of the President's belief in simple answers was that I could relax as a speechwriter, I saw, the most important was that I could relax as a citizen. The ideal of democracy wasn't being hollowed out after all. It was being reaffirmed. Bigger government, higher taxes, détente with the Soviet Union—broadly speaking, these had been the policies of experts. Smaller government, lower taxes, and peace through strength—broadly speaking, these had been the policies of ordinary Americans. In championing these simple, commonsense policies, Reagan had returned ordinary Americans to the center of national life. Expert knowledge has its place, needless to say—Reagan himself relied on experts to advise him on tax policy, foreign affairs, defense, and every other aspect of administration policy. Yet even in late-twentieth-century America, a nation in which life had become as intricate and interdependent as John George Kemeny claimed, the insights of ordinary Americans remained just as valid as those of anybody else.

Just how qualified do you have to be to form your own opinion on the issues, cast your vote responsibly, or run for office? You need no qualification, I saw, but citizenship. Anybody can mouth off on tax rates or foreign affairs, of course. But should the rest of us listen when he does? Yes, we should. His opinion is likely to be just as good as that of anybody else.

—— *My Desk* ——

RONALD REAGAN, at the big desk in the master bedroom. His example means still more to me today than it did while I was working for him. You see, it has been a long time since I read a briefing book prepared by the CIA, listened to Tony Dolan expound on the metaphysics of the Cold War, or participated in a lunchtime debate in the White House mess about whether the Soviets would pull out of Afghanistan. But if Reagan could keep in touch with events while he was just an ordinary citizen, so can I. I subscribe to political journals, read a lot of biography and history—on my bedside table as I write is David Kennedy's history of the United States during the Great Depression and World War II, *Freedom From Fear*—and lead the odd political conversation at the dinner table. Not, I admit, that I can take the conversation very far now that the children are still too young to know the difference between the powers of the President and those of the governor of California. But still. I'm getting them in training.

Our history and ideals are accessible to anyone—that, I realize now, was one of Ronald Reagan's most important insights. For a Chinese to acquaint himself with the story of his nation, he'd have to read up on four thousand years of history in which empires rose and fell and entire systems of thought and feeling were first established and then overturned. For an American to acquaint himself with the story of his nation? Easy. Once you've read the Declaration of Independence and the Constitution, you've got the essentials. With a little additional reading you can learn about the Founders, then trace their debates across a mere two centuries and into your morning newspaper. In each age, you'll recognize, the great issues have resolved themselves into questions on which any citizen was capable of forming a valid opinion. Should we separate from the mother country? Should we remain half-slave and half-free? Should we concentrate ever more power in Washington or limit the growth of government? Should we isolate ourselves or play a leading role in the world, standing up to the Nazis? The Communists? And now the Islamists?

The accessibility of the American story—the way you can almost *touch* it—represents a central portion of our inheritance, right along with self-government and the Bill of Rights. Ronald Reagan enabled me to see that.

This may be the shortest chapter in the book—there is only so much I can tell you, after all, about a man at a desk—but the lesson it contains is one of the most valuable. You don't have to be an expert to participate in American life. All you have to do is read some history, follow the news—and think for yourself. You have a head. Use it.

—————— *Six* ——————

THE MAN
WITH THE
NATURAL SWING

Easy Does It

When I stopped by his office to say hello this afternoon, David Gerson showed me one of his favorite mementoes from the Reagan years, a big photograph. Taken during the first term, the photograph shows President Reagan, Vice President Bush, and Treasury Secretary Donald Regan on the practice tee at the famous course in Augusta, Georgia. (The photograph was a gift from Regan. David, who worked in the White House, had helped Regan coordinate policy between the White House and the Treasury.) An instant earlier, the President and Vice President had teed off, and the camera had captured each man at the end of his swing, club in the air, eyes following the ball.

Although George Bush had been a golfer all his life, David explained, Ronald Reagan had never taken the sport seriously. "But look at them," David said, "and ask yourself, 'Who has the easiest, most natu-

135

ral swing?' The answer is Reagan, no question. Bush is just trying too hard."

David keeps the photograph where he can see it from his desk. As the number two man at the American Enterprise Institute, he holds a big, demanding job, and whenever he starts to get tense he looks at the photograph for a moment, reminding himself to loosen up. "I think of that picture as a life lesson," David said.

You MAY RECALL a scene from the movie *Animal House*, probably the funniest college comedy ever made. Escaping the noise of a party taking place below, four students, a man and three women, have seated themselves on a fraternity house staircase. The man looks intense and intellectual. Accompanying himself on a guitar, he sings the ballad "The Riddle Song" while the women listen soulfully. "I gave my love a cherry that had no stone," the singer earnestly intones. "I gave my love a chicken that had no bone." Descending the staircase, a fat, unkempt fraternity brother named Bluto (played, immortally, by John Belushi) happens upon the scene. Bluto stops, cocks his head, and listens with a puzzled expression, as if unable to understand why the singer is comporting himself with such seriousness. Suddenly Bluto snatches the guitar, and, grasping the instrument by the frets, smashes it against the wall. Bluto hands what remains of the guitar back to the young man, shrugs, and continues down the staircase.

Animal House was released while I was at Dartmouth, and my buddies and I must have trooped down Main Street in Hanover, New Hampshire to the Nugget Theater to see the movie at least half a dozen times. (One of the screenwriters, Chris Miller, himself a recent Dartmouth grad, had stated in interviews that he'd based much of the movie on his experiences at the college. That egged us on.) Every time we saw Bluto smash the guitar, my buddies and I went wild, cheering and applauding and stomping our feet. Instinctively, we understood that Bluto and the intense young man represented two opposing approaches to life. Probably pre-med or pre-law, the intense young man was high-minded, serious, repressed, and hardworking, the kind of student who would never even have shown up at a fraternity party unless he'd already completed all his assignments for the weekend. Bluto was vulgar, carefree, spontaneous, and irresponsible, the kind of student who could always be counted on as the life of a party, even if all that stood between him and academic probation was the studying he should have been doing while he was smashing guitars. The intense young man lived in his mind, devoting his attention to worthy causes. Bluto lived in his body, devoting his attention to women, sports, pizza, and beer. The intense young man seemed contrived. Bluto seemed natural. Bluto was our man. We wanted to be just like him ourselves.

The problem was that we couldn't. A lot of my college buddies were pre-med or pre-law, just like the intense young man, and even those of us who had no plans for graduate school nevertheless found ourselves in an educational institution in which getting decent grades required a lot of work. If we modeled ourselves on the intense young man, we'd hate ourselves. That would be bad. But if we modeled ourselves on Bluto, we'd flunk out. That would be worse. Much as we cheered Bluto, then, we knew we could adopt his approach to life only within certain limits. Bluto versus the intense young man, the serious versus the lighthearted. The trick lay in striking a balance.

Now, a lot of my classmates—especially those who had gone

to prep schools or whose parents had attended Ivy League insti-tutions themselves—seemed able to strike a balance with no problem, even with a certain style. Somehow they knew when to quit studying on Friday evening, how long to ski on Saturday afternoon, how late to party on Saturday night, and how to cure their hangovers on Sunday morning so they could get back to their books by Sunday afternoon.

Me? In all four years of college I never even began to get the balance right. I'd gone to a public high school, not a prep school, and I was the first member of my family to attend an Ivy League institution. My background, I figured, meant I had to work harder than other students—if I did badly at Dartmouth, I'd have felt as though I'd let down the whole town of Vestal, New York—so I'd do just that, burying myself in the library to study until all hours of the night, behaving, my pre-med friends told me, like a classic Type A personality. After ten or twenty days, though, something would snap. I'd suddenly find myself physically inca-pable of sitting still at a desk—sometimes even of opening a book. Once that happened, I'd quit studying and go on a binge. The binge might involve alcohol or it might not—well, not much, anyway. But it would always involve pursuits that made me feel a kind of wild, delirious abandon. Sessions of poker and cigars that ran for hours. Weekend road trips to women's colleges. Fraternity parties that made those in *Animal House* look reserved. One night we flooded the basement of the Tri-Kap fraternity house with beer to a depth of more than an inch, then took turns running, leaping, and hydroplaning across the floor on our bellies. When the binge ended—I'd wake up one morning, my head pounding, and realize I'd had enough—I'd grab my books and head off to the library to begin studying like a classic Type A per-sonality all over again. First I'd be the intense young man. Then

I'd be Bluto. One or the other, this or that. At Oxford, where I studied for two years after graduating from Dartmouth, I followed the same pattern, working too hard, then playing too hard. Strike a balance? Never.

This brings us to the White House. If I'd been unable to figure out how to deal with my studies, I was conscious when I arrived at the White House from Oxford, I was going to need help in figuring out how to deal with life. So I studied a man who had it down.

—— *How to Handle Ray Charles* ——

M Y FATHER was a very physical person," Ron Reagan says. "While he was never big or strong or fast enough to be a professional athlete, there was this just innate athleticism to him. He was at ease with his own body."

At ease with his own body.

Strange though this may sound, when I arrived at the White House the proper place of the body in the life of the grownup puzzled me. During my college binges, my body had been my master—a jolly master, but a master—driving me to wolf down pizza or guzzle beer. Then, during my Type A phases, my body had become my prisoner, locked away in an attic tower while I pretended I was some kind of brain. The idea that a person could be at ease with his body never entered my head until I encountered the fortieth chief executive.

Every time the President held a meeting in the Oval Office, stood at the head of a receiving line shaking hands, or walked down the aisle of Air Force One greeting the passengers, one of the White House photographers would record the event, making sure to get at least one shot of the President with each of those present. Each time I was photographed with the President myself, I did what every member of the staff did, going upstairs to the photographers' office in the Old Executive Office Building to look over the contact sheets, then choose a negative or two that I'd like to have printed. (My parents never quite got used to the idea that I worked for the President, so I had to keep sending them proof.) After I'd done this a few times, it struck me: In each of the dozens of photographs I'd examined, Ronald Reagan had stood out.

The President wasn't always the best-dressed person in a photograph; he sometimes wore a blue suit with a plaid so big it looked like a television test pattern. He wasn't always the most handsome person in a photograph, either; something about the White House seemed to attract those who were unusually good-looking, and although we speechwriters tended to look just the way writers everywhere tend to look, which is to say like riffraff, some members of the staff were, literally, beautiful people. Yet Ronald Reagan was always the most—well, I'm not even sure how to put it. At ease. Composed. Physically present. One of the photographers told me it was almost impossible to take a bad picture of the President even if you tried, which, to break up the monotony of his work, he'd sometimes done. Reagan always exuded a certain physical lightness or grace.

To some extent, of course, the President's physical attributes represented a lucky accident of birth. A famous photograph of Reagan as a lifeguard at Lowell Park, north of Dixon, Illinois, while he was still only sixteen, shows a young man who was trim, broad-shouldered, and strikingly attractive—if the Reagan of that photograph had worked in the White House he'd have numbered among the beautiful people on the staff himself. Yet

Reagan never took the advantages his genes had conferred on him for granted. He watched what he ate. He got plenty of sleep. "He always said, 'I need eight hours of sleep and I do better on nine,' " Lyn Nofziger recalls. And he exercised constantly. After the attempt on his life, for instance, Reagan had a set of weights installed in the residence, then worked out with them so diligently each afternoon that he quickly added more than an inch to the diameter of his chest. And when he visited the ranch, he'd perform hard physical labor for hours at a time, just as we've seen, clearing trails, pruning trees, or setting fenceposts. "Sometimes he'd work so hard," says Dennis LeBlanc, who worked alongside the President, "that the Secret Service guys would start to worry he'd get dehydrated." Unable to persuade the President to rest for his own sake, the agents would ask LeBlanc to get the President to rest by knocking off for a while himself. After Reagan had been at the ranch for a couple of weeks, LeBlanc says, "He'd come out one morning and he'd say, 'Boy, I was able to pull in a notch on my belt on this trip.' "

Reagan not only took good care of his body, he made sure he looked good. Over the years, I gradually realized, he'd picked up a set of techniques for enhancing his physical appearance. When he posed for photographs, for example, Reagan would drop his hands to his sides, then touch the tips of his fingers to the tips of his thumbs, a practice he'd adopted in Hollywood. "It doesn't feel natural," I once heard him explain, "but it looks that way." To keep from appearing rumpled, he wore ventless jackets, cut to fit snugly. And to offset his relatively small head, he tied big knots in his neckties, a lesson he'd apparently learned from the Warner Brothers costume department. Sometimes Reagan's techniques proved almost *too* effective. When the President and Vice President posed together for their official portrait, one of the

White House photographers told me, Reagan looked so utterly immaculate that he made Bush seem disheveled. To get the Vice President's pant legs to look half as neat as the President's, which had such sharp creases they might have sliced cake, the photographer had to hang binder clips from both of the Vice President's cuffs.

Now when I first thought about it, there seemed something a little odd in all this. More than most people, Ronald Reagan lived in his mind. He would never have felt at home hobnobbing about Hegel or particle physics in the faculty lounge of a major university, I'll grant you. Yet all the same, in the strict sense of the word he was an intellectual. He read widely and wrote constantly, just as we've seen. And he took his conclusions so seriously that he organized his entire life around a few central propositions, including the importance of liberty, the dynamism of free markets, and the fundamental wisdom of the American people. Much as he valued ideas, however, there was never anything about him of the bow-tie–wearing, absentminded professor. He was always too solid, too completely *here*. For a man who lived in his mind, Reagan spent a lot of time in his body. How, I wondered, did Reagan do it?

"The mind *or* the body? You mean we have to choose? My friend, you sound like a Manichean," Father Albacete said. Over a dinner, this time of Big Macs and French fries, I'd told the priest that Reagan's physical grace struck me as curious. Father Albacete had replied with a dose of intellectual history, describing Manichaeism, the third-century philosophy that saw mind or spirit as intrinsically good, matter as intrinsically evil. "The Manicheans," he said, "thought we were all little sparks of the divine trapped in big globs of mud or some crazy thing." Late in the third century, the church condemned Manichaeism, pronouncing it a heresy. Yet even when Constantine made Christianity the official religion of the Roman empire early in the fourth century, Manichaeism proved difficult to stamp out.

"It wasn't just a philosophical system, it was a turn of thought, a way of looking at the world," Father Albacete said. Even today, when scarcely anyone had even heard of it, Manichaeism was all around us. "Who hasn't stood on the scale, realized he was overweight, and felt trapped inside his own body? Anorexia, bulimia—they're all an effort to reject the physical world. Or look at booze and drugs. What are they but a way of fleeing the physical world for realms of pure thought and feeling? St. Thomas Aquinas tells us that Jesus instituted the sacrament in the form of bread and wine—in the form of ordinary, physical stuff—for just this reason. We're so pigheaded, we have to be *forced* to see that the physical world is good."

Reagan, Father Albacete suggested, understood the goodness of the physical world. "He has what the theologians would term 'integrity,' meaning a oneness or unity of being. He's not mind *or* body. He's mind *and* body, integrated, together, whole." Granting his body its place, Reagan treated it with respect, taking pains to care for it—even taking pains to show it off to its best advantage. He saw his body as neither his master nor his prisoner, but as a sort of friend.

Not long after my conversation with Father Albacete, I witnessed an incident that I still consider the best illustration of Reagan's physical grace or composure. At the taping of a 1983 television program marking the twenty-fifth anniversary of the Country Music Association, all the performers lined up across the stage of Constitution Hall, an auditorium near the White House, while the President and Vice President made their way down the line, shaking each artist's hand. Willie Nelson, I noticed from my seat in the audience, happened to be standing next to Ray Charles. As Reagan and Bush approached, Nelson whispered in the blind musician's ear, evidently telling Charles that he was about to meet both members of the executive branch. Charles broke into a grin. Then he began swaying from side to side with excitement. When the Vice President reached Charles, Bush did just what I'd have done, fumbling for Charles's hand to give him a simple handshake. The Vice President grasped Charles's hand for a moment or two longer than he had those of the other performers, but even so the gesture seemed insufficient, a puny return on the enthusiasm Charles himself was displaying as he rocked and swayed. When the President reached Charles, he, too, took Charles's hand. But as Charles continued to rock and sway, his movements as stiff and erratic as those of an autistic child, Reagan pulled the musician to his chest. Then Reagan placed his arms around Charles, enfolding him in an embrace.

Ronald Reagan, at ease with his own body.

—— *Outputs* ——

A TYPICAL DAY in the life of the fortieth chief executive: President Reagan would arise between seven and seven-thirty. Over breakfast with Mrs. Reagan at seven forty-five, he would read the White House news summary, a compilation drawn from newspapers, radio, and television. At nine o'clock he would reach the Oval Office, beginning a formal working day of between five and eight hours. The President would hold meetings, place telephone calls, discuss business over lunch (he and the Vice President lunched together every Thursday), step into the press room to make an announcement, stroll over to the East

Room or out to the Rose Garden to deliver a speech, and devote an hour or more to what appeared in his published schedule as "private time," time during which he might read, write letters, work on speeches, or slip back upstairs to the residence for a nap. (After the President's surgery for colon cancer, I made a comment to a Secret Service agent about Reagan's schedule, saying it seemed remarkable that a man in his mid-seventies could remain so active. "Yeah," the agent replied, "but he gets more rest than you might think.") Throughout the day the President would turn repeatedly to the printed schedule he kept on his desk, methodically drawing a line through each item as soon as he had completed it. And he would make a point of remaining on time. "The only time I ever remember him so angry he threw his glasses," Ed Meese says, "was when he was overscheduled. He just hated it when people were kept waiting. He felt it was discourteous."

When he left the Oval Office between four-thirty and five-thirty, the President would spend about half an hour working out in the small gymnasium he'd had installed in the residence. Afterward, he might accompany Mrs. Reagan to a dinner at an embassy or hotel or join her for a reception at the White House itself. Otherwise he and Mrs. Reagan would have dinner alone in the family quarters, often while watching television. Later the President would spend an hour or two signing documents, editing speeches, or going over his briefing books. Sometime during the evening he would write in his diary, composing a few paragraphs in which he took the measure of the day. Between ten and eleven, he would go to bed. Usually he would read a book before turning out the light. By all accounts, he would sleep soundly.

Methodical. Well-paced. Unharried. Ronald Reagan never broke a sweat.

"He never confuses inputs with outputs," Clark Judge, my

colleague on the speechwriting staff, once remarked. Clark held a degree from Harvard Business School, so when he talked about inputs and outputs, we listened. "People talk about how many hours a day Nixon and Carter put in," Clark said. Richard Nixon used to work at all hours—long after midnight during one of the mass protests against the Vietnam War, Nixon instructed the Secret Service to drive him to the Lincoln Memorial, where he emerged from his limousine to debate war aims with a startled group of hippies—and Jimmy Carter was known to put in one sixteen-hour day after another. "Who *cares* how many hours a day a President puts in? It's what a President *accomplishes* that matters."

Reagan, Clark argued, concentrated his attention on the few tasks he alone could perform. He set overall administration policy, made two or three critical decisions a day, and gave speeches in which he explained his goals to the American people. He left everything else to the staff. That gave him plenty of time to exercise, get his sleep, and enjoy himself, activities that in turn enabled him to remain fresh and composed in performing his duties as President. "It works, right?" Clark said. "I mean, compare Reagan's outputs—what he's actually accomplishing—with Nixon's and Carter's. Reagan just blows Nixon and Carter away."

Type A versus non–Type A. When I arrived at the White House, as I've explained, I had the notion that I'd have to drive myself to accomplish anything in life. Yes, I know. The fable of the tortoise and the hare dates back 2,600 years. But I could never buy the moral. Slow and steady wins the race? Losers *would* say that. Ronald Reagan changed my thinking. He had begun life without any special advantages, yet instead of driving himself he had adopted a methodical, unharried approach to his work—even as a young actor, still building his career, he always made a point of getting plenty of rest and recreation. Yet just look at what he accomplished.

It's a lesson I don't suppose I'd have predicted anyone would ever learn from a man who had climbed to the pinnacle of American politics, but I learned it from Ronald Reagan all the same. Drop the Type A act. *Relax.*

——*Take-a My Heart* ——

EXAMINE THE *Public Papers of the Presidents,* the volumes published by the Government Printing Office that contain all the utterances of our chief executives, and you'll see that during his eight years in office Ronald Reagan produced enough quips, jokes, stories, and witticisms to fill a couple of volumes in their own right. When during the President's 1984 reelection campaign a reporter shouted, "What about Mondale's charges?" Reagan replied simply, "He ought to pay them." Poking fun at his age at a dinner for news photographers, Reagan said he looked young because he had a good makeup team. "It's the same people who've been repairing the Statue of Liberty." Looking back on all the meetings with Reagan that I attended—including meetings for no purpose other than to take a group photograph of the speechwriting staff—I'm unable to recall a single instance in which the President failed to tell at least one joke.

Now, when I arrived at the White House, I tended to think of humor as something that happened at fraternity parties, on road trips, or during visits to the Nugget Theater to cheer Bluto in *Animal House.* Humor was low. It was coarse. The best that could be said of humor was that it represented a release from the real business of life. Humor, if I may put it this way, wasn't *serious.* Reagan taught me otherwise. He employed humor in what may be the most serious endeavor to which a chief executive can

147

address himself, winning and sustaining the support of the American people.

In differing circumstances the President would employ humor to differing ends. Sometimes he would use humor to establish a bond between himself and an audience, doing so with such deftness that even we speechwriters would find ourselves astonished. In the summer of 1986, for example, I was assigned to write a speech for the President to deliver to the Knights of Columbus, a national organization of Catholic men. As I worked, devoting much of the speech to the importance of charity—the Knights of Columbus raised tens of millions each year to support schools and hospitals—one problem kept bothering me. Many of those in the audience would be Irish Catholics. Ronald Reagan was an Irish Protestant. The Irishmen in the audience wouldn't hold Reagan's faith against him, exactly—polls showed that the President enjoyed the support of a majority of Irish Catholics—but it would place a small, unspoken gap between the audience and the President all the same. I wracked my brains. How could I bridge that gap? Aside from including a passage in which Reagan announced plans to convert, I could think of nothing. Once I'd completed the draft, I did my best to put the problem out of my mind.

On the day of the speech, I turned the television set in my office to the closed-circuit White House channel, then settled myself in one of my big armchairs to watch the President deliver the speech from the Oval Office. Reagan spoke the opening few lines just as I'd written them. Then he paused. "We're trying out a new technology today," he said, referring to the small speaker that sat atop his desk, "one with a hookup that will enable me to hear you . . . if you laugh or applaud. And I thought the best way to test it would be to tell an old story." This wasn't part of my text.

The President's eyes twinkled. I leaned toward the television. What was he doing?

"It has to do with a young fellow," the President said, "that arrived in New York Harbor from Ireland."

And a short time later, he started across one of those busy New York streets against the light. And one of New York's finest, a big policeman, grabbed him and said, "Where do you think you're going?" [The President delivered the dialogue in a wildly exaggerated brogue.]

"Well," he says, "I'm only trying to get to the other side of the street there."

Well, when that New York policeman, Irish himself, heard that brogue, "Well," he said, "now, lad, wait." He says, "You stay here until the light turns green, and then you go to the other side of the street."

"Ah," [the young fellow] says, "the light turns *green.*"

Well, the light turned orange for just a few seconds, as it does, and then turned green, and [the young fellow] started out across the street. He got about fifteen feet out and he turned around, and he says, "They don't give them Protestants much time, do they?"

The audience howled with laughter, then broke into applause. With one joke, the President had established his own Irishness, acknowledged that whereas his audience was Catholic, he himself was Protestant, and then made light of this gap between them in a way that enabled him and the audience to share a moment of pure delight. It was perfect. I mean *perfect.*

Sometimes the President would use humor to make a point. Marking up a speech I'd written soon after I'd joined his staff, for example, the President placed a note in the margin next to a passage about the importance of political participation: "Insert Farmer Joke."

"What's the farmer joke?" I asked Dana Rohrabacher, who of all the speechwriters had known the President longest.

"*Which* farmer joke?" Dana replied. "He has a hundred of them."

This particular farmer joke, as I learned when I listened to the President deliver the speech, concerned a rustic who spent weeks working on an overgrown creek bottom, clearing the land, fertilizing it, planting it with vegetables, and then tending it as the crops came in. Proud of all he'd accomplished, the farmer invited the preacher out to take a look one Sunday after church.

"Those are the biggest, reddest tomatoes I've seen in my life," the preacher said. "Praise the Lord! And that corn! Those are the most enormous ears I've ever set eyes on. God be thanked!" The preacher, the President explained, continued in this way, admiring the crops and praising the Lord, until the farmer broke in. "Preacher," the farmer said, "you should have seen this creek bottom when the Lord was taking care of it by Himself."

If a picture is worth a thousand words, so is a punch line as sweet as that. I could have written page after page of high-flown oratory. None of it would have made the President's point as vividly.

Not long after I heard the President tell the farmer joke, I heard him interrupt a speech about the failures of Communism to ad-lib another. The joke, as the President told the audience, was a favorite inside the Soviet Union itself. It concerned a Soviet housewife who needed a new refrigerator. When she placed her order, she was told it wouldn't be filled for ten years.

"Will you be delivering the refrigerator in the morning or the afternoon?" she asked.

"Comrade," the refrigerator salesman replied, "as I just told you, it won't be for *ten years*. Why would you possibly want to know whether we'll be delivering the refrigerator in the morning or the afternoon?"

"Because," the housewife replied, "the plumber is coming in the morning."

If there was a more effective put-down of Communism, I never heard it.

Making a point, achieving a bond with an audience—whatever the particular use to which Reagan might put a quip or a joke, his humor, I gradually realized, always produced a larger and more important effect as well. It reassured people. It made them feel better. When Reagan told a joke, he wasn't trying to make an audience like him, but giving it a gift. He was reaching into that zone deep within himself that was somehow always filled with delight—just picture the sparkle in his eyes as he anticipated the ripple of laughter he knew a good punch line would elicit—to give his listeners a little delight of their own. Humor, he saw, provided a kind of universal solvent, capable of washing cares away, and he always took the time to make people feel better. Reagan never lost his sense of humor—not when the economy went sour, not during the lowest moments in our relations with the Soviets. He often claimed the country was bigger than its problems. When they heard him making jokes, Americans could see he believed it.

Which brings me to a second matter. Ronald Reagan appreciated humor not only because he could employ it to political effect. He appreciated humor for its own sake. If you'd wanted proof, you'd only have had to conduct a survey of the White House staff, asking people to relate jokes the President had told them in private, where his humor couldn't do him any political good. You'd have come up with dozens, and quite a few would have been so politically incorrect that if they'd ever leaked they'd have gotten the chief executive into trouble. One afternoon while still on the Bush staff, for instance, I had a meeting with the Vice President on a Thursday, the day he always had lunch with the President. I couldn't get the Vice President to settle down and talk about speeches until he'd let me in on the joke the President had just told him.

The Pope, it seems, needed a heart transplant, but, like a lot of people in that position, was finding it difficult to locate a donor. "No problem, Your-a Holiness," Cardinal Casaroli, the Vatican secretary of state, who was known to both the President and the Vice President, said. "You just-a listen to me. I'm-a gonna tell you what we're-a gonna do." ("You have to imagine the President telling the story," Bush said. "My Italian accent isn't all that good. His is just flawless.") Cardinal Casaroli then disclosed his plan to the Pope.

The next morning the Vatican newspaper, *L' Osservatore Romano*, acting on Casaroli's instructions, broke the story about the Pope's condition, announcing that John Paul needed a heart donor immediately. By noon, half a million Italians had jammed St. Peter's Square, all of them expressing their eagerness to serve as heart donors by shouting up at the papal balcony, "Papa! Papa! Take-a my heart! Take-a my heart, Papa!" Overwhelmed by this display of devotion, the Pope stepped onto the balcony to acknowledge the cheers. Then, returning inside, he said, "There's a problem, Casaroli. How do I choose one person from among so many?" "Don't worry, Your-a Holiness," Cardinal Casaroli said once again. The cardinal lifted a pillow from a chair and began to plump it.

A moment later, the Pope and Cardinal Casaroli stepped onto the papal balcony together. "In one-a minute," Casaroli announced to the crowd, "the Holy-a Father, he's-a gonna drop a feather from this-a balcony. And whoever that feather-a lands on, that person is-a gonna have the honor to give the Holy-a Father his heart." The crowd cheered. The Pope released the feather. And as the papal feather floated down to St. Peter's Square, half a million Italians shouted, "Papa! Papa! Take-a my heart!" *Pfft! Pfft!* (Bush pursed his lips and blew as if trying to keep a feather from landing on him.) "Take-a my heart, Papa!" *Pfft! Pfft!*

See what I mean? Ronald Reagan might have retained the support of the Irishmen in the Knights of Columbus if that one had ever gotten out, but he'd have found himself in a lot of trouble with the Italians.

Reagan knew perfectly well that life often proves tragic, of course. Even if you'd never known that his father was an alcoholic, you'd have sensed that Reagan himself had suffered when you saw the empathy he conveyed to the bereaved at memorial services, such as the service for the Marines killed in Lebanon or that for the servicemen who died in Grenada. Ultimately, though, Reagan considered human history a comedy—not a trifle or an absurdity, but a solemn story that would end, finally, in happiness, because God was the author, and God was good, and that was the kind of story He'd write. Reagan taught me to appreciate the uses of humor, as I've said. But he also taught me to appreciate the *meaning* of humor. The world contains more good than bad, more courage than cowardice, and more reason for smiles than for tears. Laughter is a profession of faith.

—— *How Angels Fly* ——

JOURNAL ENTRY, MAY 2002:

Hardly anybody ever saw the ranch but the Reagans, the couple of men who helped the President with the work he was always doing up there, the odd reporter Reagan invited up for an interview, and a few dozen Secret Service agents—even the President's chiefs of staff and national security advisers almost never went up, instead remaining at staff headquarters at the Los Angeles Biltmore. The ranch was Reagan's haven. He could do with it as he pleased. And doing what he pleased, as I learned when I toured the place today, included indulging his sense of humor.

In a couple of trees near the entrance, Reagan stuck fake owls. On the front door, he put up a plaque that reads, "On this site in 1897, nothing happened." In the tiny shower stall just off the master bedroom, he installed a shower head in the shape of the Liberty Bell. And on the wall across from the small bar he set up at one end of the kitchen, he mounted the heads of two jackrabbits, one of which had sharp canine teeth and antlers. "Male and female jackalopes," Marilyn, my guide, explained, chuckling. "A cross between jackrabbits and antelopes. He used to show them to reporters. Apparently he was quite often able to persuade reporters from back East that they were real."

But the crowning glory, the gag he liked so much that it must have taken him half a day to set it in place? It was a structure that dominated the view from the Reagans' bedroom, and I couldn't help wondering what Mrs. Reagan must have made of it. For that matter, I couldn't help wondering what the few exalted visitors to the ranch—Queen Elizabeth and Prince Philip, for example, or Mikhail and Raisa Gorbachev—must have made of it. Just ten yards from the house, Reagan had erected a shed with a half moon on the door.

When I saw that fake outhouse, my affection for Ronald Reagan grew even deeper. As G. K. Chesterton once wrote. "Angels can fly because they can take themselves lightly."

R ONALD REAGAN was a natural. A lot came to him easily—his athleticism, for instance, seems to have been inborn—but that's not what I mean here. I mean that Reagan was

154

entirely accepting of and thoroughly grounded in human nature itself. Lots of people wage war on their bodies all their lives, trying to lose weight or undergoing round after round of plastic surgery. Others try to escape from their bodies altogether, abusing drugs and alcohol, just as Father Albacete pointed out. Reagan accepted his body—accepted it, appreciated it, cared for it. Time? How many people see the clock as their enemy—how much had I struggled against the relentlessness of time myself—working more and more, sleeping less and less, constantly harried? Reagan worked hard, but he paced himself. He never strained against time. He worked within it.

You could see Reagan's acceptance of human nature most clearly of all in his love of humor. We humans are utterly ridiculous, of course—half body, half spirit, always trying to take ourselves seriously, surprised when we find the little lectures on our own importance that we're always giving ourselves interrupted by our need to use the bathroom. Instead of trying to remake human nature, as did, for example, Lenin and Hitler, Reagan simply told jokes—told jokes, then threw back his head and laughed. We're all in this predicament together, the twinkle in his eyes seemed to say, but it'll come out all right in the end.

One of my buddies at Dartmouth, a pre-med student, used to warn me that if I kept behaving like a Type A personality I'd become an excellent candidate for a heart attack in my early forties. I figured he was probably right. But what could I do about it? Turn myself into Bluto? As it was, my buddy's warnings only gave me something else to obsess on. The Gipper showed me the way out of the dilemma. To lighten my mood, I got my hands on an old stereo and moved it into my office in the Old Executive Office Building, then played Louis Armstrong, Ella Fitzgerald, and Oscar Peterson when I was doing background reading. And since Reagan exercised regularly, I began doing the same myself, jogging most mornings in Rock Creek Park and working out at lunchtime a couple of days a week in the gymnasium the Secret Service used. As a mathematical proposition, of course, each hour I spent on my body was an hour taken away from something else, but it never seemed that way. Feeling more alert and energetic, I was able to work more smoothly and effectively. My days seemed to expand, not shrink.

Now that I'm a father, I find myself thinking about human nature all the time. You see, I'd pictured myself as Fred MacMurray on the old television show *My Three Sons.* Wearing a sweater and cradling a pipe in one hand, I'd imagined, I'd dispense warm, sensible advice to children who were courteous, clean, well-groomed, and attentive. Each of my children, I seem to have supposed, would be at least nine years old at birth. Instead? Baths to give. Spills to clean up. Diapers to change. Reagan used to quip that the federal government was just like an infant, with an endless appetite at one end and no sense of responsibility at the other. I've developed a whole new appreciation for that line, believe me. Although my wife has always provided most of the care for our children, I've done my best to pitch right in, taking human nature as it is, and I'm a better father for the effort. There's no place like the changing table for making eye contact with a baby.

You can accomplish a lot more by cooperating with human nature, I learned from the fortieth President, than by straining against it. Life lasts a good long time. Easy does it.

—— *Seven* ——

WITHOUT HER, NO PLACE

Marriage Can Save Your Life

When you aren't there I'm no place, just lost in time & space.
—*Ronald Reagan in a letter to his wife, March 4, 1983*

WHEN I BEGAN this book, I dug out the journals I'd kept at the White House. I expected to find that I'd composed portraits of all the historic figures I'd observed and written probing little essays on the economics, foreign affairs, and politics of the day. What a treasure, I thought. It even occurred to me that half a century from now my children might hand the journals over to the Smithsonian, giving the nation a sprightly account of the 1980s while earning themselves a tax break.

Then I got around to reading the journals. No more than a

third of the many hundreds of pages dealt with the kind of material I'd expected to find. The other two-thirds? Scenes from the self-absorbed existence of a bachelor.

Out-of-town college buddies show up, go out drinking with me, crash overnight on my floor, then disappear. Josh Gilder and I spend hours together. Both single, we carouse, we go to movies, we throw parties, we attend parties, I drink beer, he drinks wine. One winter, Josh and I decide to get into shape. We join a health club, play squash with a couple of friends almost every evening after work, then go to a restaurant around the corner that offers all-you-can-eat tacos for $3.95. My weight, which I record every day, goes up—and, my journal records, I'm *surprised.* The following summer Josh and I spend almost every weekend sailing on Chesapeake Bay in a rented catamaran. Although Josh, who grew up sailing on Long Island Sound, knew what he was doing, I had never sailed in my life. I note in my journal that if Josh ever fell overboard—and when he hiked out over the side of the boat, that was a possibility—I'd only have been able to jump up and down, waving for help to passing container ships as the catamaran drifted out to sea. This seems to have struck Josh and me as hilarious.

Women? Yes, women. My journals are full of them. But you'd be mistaken if you supposed I cut a figure in the singles bars of Washington. I was less of a Lothario than an ornithologist. In my entries about women, I'm constantly commenting on their amazing variety, astonishing plumage, and peculiar group behavior. I often commit to my journal fragments of conversations I'd had with women, doing so with the same detachment I'd have used if attempting to establish whether they were capable of higher forms of communication. After reading my journals, you'd feel you knew Josh Gilder, Clark Judge, and half a dozen of my other male friends. About all you'd feel you knew about the women who appear was how often they turned me down for dates. I had trouble seeing women as human, let alone establishing relationships with them.

I suffered, I see now, from a couple of particular disadvantages. Growing up, I had a brother but no sisters. Then, when I went away to school, I spent four years at Dartmouth when the college, making a transition from single-sex education to coeducation, was still predominantly male—the ratio in my class was three men to every woman—followed by two years at Christ Church, an Oxford college that remained, in my first year, all male, and then, in my second, broke with five centuries of tradition by admitting, among a couple of hundred men, a couple of dozen women. I felt the allure of women—my hormones were as active as those of any young male. But huddling among men all my life, I saw women as entirely alien. My mother? Well, she was Mom. It never occurred to me that anyone might be like her.

An incident comes to mind here. Each time my Dartmouth buddies and I made a road trip to Smith, the women's college from which, as it happens, Nancy Reagan graduated, we'd begin with the same ritual, clambering out of the car to race to Paradise Pond, the ornamental lake on the edge of the Smith campus. Under cover of darkness—for some reason, the idea of making the two-hour drive never seemed to occur to any of us until seven or eight in the evening—we'd turn around, and, as we put it, "drop trow," mooning the pond. Mooning the pond. A sophomoric gesture to you, perhaps, but a sweet memory, even now, to me. Bluto would have been proud of us. One night my roommate reached the pond ahead of the rest of us. He turned around, dropped his trousers—and then realized that a couple of Smith women were sitting under a tree where they could see him. For the rest of the night all of us worried that the women had gotten a good enough look at his face to be able to point him out to their friends if they spotted us at a party. Driving to Smith, we'd been in high spirits, singing one rowdy song after another because we'd be meeting some women. The moment we had, the women spoiled all our fun.

For my first couple of years in the White House, my journals indicate, my bachelor existence suited me just fine. I had a girlfriend I'd met at Oxford, but she was English, and, apart from making a couple of visits to Washington, remained on the other side of the ocean. That meant I was free to go right on spending almost all my time among men. The first floor in the Old Executive Office Building might as well have been laid out for the purpose. On one side of the hallway at the southern end of the building lay the offices of the speechwriters, most of whom were men—of the fourteen people who served as speechwriters during Reagan's years in office, only two were women—while on the other side of the hallway lay the offices of the researchers, all of whom were women. Spending time on one side of the hallway was like being back at college. Walking across to the other side was like going on a road trip. Not, of course, that I ever indulged in any mooning. But I could flirt for awhile, then leave. Half an hour a week on the telephone to England and a trip across the hallway every couple of days. Why would I ever want anything more to do with women than that?

This brings me to Nancy Reagan.

—— *Tales of the First Lady* ——

TWO TALES of the first lady:

The first took place when I traveled with Mrs. Reagan one day in 1984. Her staff had asked the speechwriting office for help with her speeches, and I was tagging along now to listen to her speak, getting a feel for her style and pacing so I'd be able to write for her myself. All day the three members of the first lady's staff who were traveling with her—her chief of staff and two assistants, people who always seemed relaxed and pleasant back at the White House—had struck me as manic. Would the lights at each podium be just right, illuminating Mrs. Reagan with a gentle glow, not a glare? The first lady was very particular about lights. Had the sound systems been tested? What about the microphone? Had it been set at the right height? The first lady would expect the microphone to be high enough to amplify her voice but low enough to remain out of the line of sight when she was photographed. Were we remaining on schedule? The first lady was a stickler for punctuality. As we returned to Washington aboard one of the small jets the Air Force put at the first lady's disposal, the steward served a chicken dinner. Mrs. Reagan's staff and I, seated in the rear cabin, found the dinner so delicious that we asked the steward how he had prepared it. As we were talking with him, the first lady summoned her chief of staff to the forward cabin. He returned a moment later, shaking his head. Mrs. Reagan, he reported, had found the chicken just as delicious as had the rest of us. The problem? Chicken had been served the last time she'd flown. Was a little variety out of the question? "There's always something," he said.

Midway through the flight, the first lady's chief of staff, who had set aside time for me to discuss speeches with Mrs. Reagan, directed me to join the first lady in the forward cabin. She was seated at a table reading *Town & Country* magazine. I sat down across from her. Putting down her magazine, she began explaining herself.

"Now," she said, an earnest look on her face, "I'm not a . . ." Mrs. Reagan completed this sentence not with words but with a gesture, making a fist of one hand, then punching the air. Next she assumed a relaxed, pleasant expression. "I'm more of a . . ." She completed this sentence by gesturing with both hands, palms down, as if smoothing the wrinkles from an invisible garment.

The first lady looked to see if I'd understood. It would have been awkward to tell her I didn't know what she was talking about, so I just nodded. Pen in hand, I prepared to take notes, hoping to figure out what she meant as she continued. She didn't continue. She only said, "Well, I'm glad we understand each other," then glanced down at her *Town & Country*, signaling that our conversation had come to an end. With nothing more to guide me than two gestures—a little fist, punching nothing, then two palms smoothing the air—I returned to the rear cabin.

When I told the first lady's chief of staff what had happened, he wasn't at all surprised. The fist? Although her husband had spent the last several decades attacking Democrats, liberals, Communists, and other threats to the republic, he said, Mrs. Reagan disliked controversy. That was where the smoothing gesture came in. Mrs. Reagan wanted her speeches to be tranquil. "She likes to tell people, 'I only wish I could give our children life all tied up in a ribbon,'" he said. "'Life all tied up with a ribbon.' Be sure to use that phrase."

Nancy Reagan, I'd decided by the time we landed, could hardly have differed from her husband more. He never worried about details. She insisted on attention to the smallest details. He concerned himself with the substance of speeches. She concerned herself only with their tone.

The second tale took place not long afterward in the Rose Garden. I was standing just behind the President one morning as he delivered remarks I'd drafted—as I recall, his audience was a group of young people, perhaps 4-H'ers or Girl Scouts. Although his performance was fluid enough, his pacing was off. He seemed perfunctory and detached. For once, I thought, Ronald Reagan was having a bad day. Then a movement on the second floor of the residence caught his eye. He glanced up. Mrs. Reagan was standing at a window. She smiled. The President beamed. She waved. He waved—then had everyone in the Rose Garden turn around to wave, too. When he returned to his remarks, the President picked up his pace, appearing more involved and energetic. Even—well, younger. A smile and a wave from his wife. They were all Reagan needed.

I offer these two tales to convey something of the confusion I felt when I first attempted to understand the Reagans' marriage. He approached life one way, she another; what, I wondered, could a man who read a political magazine such as *National Review* ever have to say to a woman who read a society magazine such as *Town & Country?* But just look at the effect she had on him that morning in the Rose Garden. She waved and smiled—and suddenly he was more *alive.* Ronald and Nancy Reagan were tight. They were a unit. And I just couldn't figure it out.

——*A Different Point of View* ——

M Y MOMENT of insight took place during a speech meeting in the Oval Office. The day before, I'd traveled with the first lady once again, this time tagging along to see how she handled the material I'd written. After one speech, the first lady's staff had arranged for Mrs. Reagan to give a ride in her limousine to a girl whose story the staff had wanted the first lady to hear; although only eleven or twelve, as one member of the staff told me, the girl had already become a drug addict, then kicked her habit, undergoing treatment. During the speech meeting in the Oval Office, I heard the story of the little girl all over again. "Why, just yesterday," the President said, "Nancy met a girl who had already been through a treatment program—and she wasn't even a teenager. Addicted as young as that. Imagine."

As the meeting continued, I found myself picturing the scene that would have taken place the evening before. After a day in which members of Congress would have tried to get him to raise taxes or spend more money, members of his own staff would have competed for his attention, trying to get him to give them this or that decision, and the press corps would have shouted questions at him, trying to get him to give them newsworthy or quotable material, Ronald Reagan would have walked from the Oval Office to the residence, taken the elevator to the family quarters, worked out, changed into comfortable clothes, then sat down to dinner with the one person he would have seen all day who wanted nothing from him, his wife. What a relief it must have been for Ronald Reagan, I saw, to lay down the burdens of the presidency to become only Nancy's attentive husband, listening as she told him about her day.

After that speech meeting, I decided that the first lady was a lot like one of those high school composition exercises in which you had to write about the same person from different points of view. Seen from the point of view of the White House staff, Mrs. Reagan appeared formidable. Seen from the point of view of her husband, she appeared devoted, protective, and loyal. Her attention to details? How many times in Ronald Reagan's political career must he have looked and sounded better when he delivered a speech because his wife had made certain someone had double-checked the lights and the sound system, appeared at his best in a press conference or debate because she had refused to permit the staff to overschedule him, or derived particular pleasure from a meal because she had planned the menu? Even her insistence on giving only the most soothing of speeches, I recognized, demonstrated her concern for her husband. He attracted enough controversy already. When she spoke in public, she wanted to adorn his administration, not cause trouble.

Point of view. That, I saw, was the secret. Consider a couple of stories about Mrs. Reagan that I found myself looking at in a new light.

One dated to Reagan's 1966 campaign for governor. Then a college sophomore, Dana Rohrabacher found himself devoting dozens of hours each week to Youth for Reagan. Unfortunately, as Dana soon learned, Youth for Reagan suffered from infighting between two factions, the Young Republicans and the Young Americans for Freedom. "It was the old story in the conservative movement," Dana had said, telling me the story. "The right couldn't stand the far right." Tired of the infighting, the Reagan campaign announced that Youth for Reagan would be disbanded.

Dana felt beside himself. He'd already organized nearly a hundred college kids to walk precincts for Reagan. How could he tell them the campaign no longer wanted them to have an organization of their own? Dana decided to have a talk with the one person he knew could reverse the campaign's decision, Ronald Reagan. How did a college kid get an appointment with a gubernatorial candidate? He didn't. He talked a friend into joining him, borrowed a couple of sleeping bags, then camped out in the candidate's back yard.

At seven the next morning, the back door of the candidate's house swung open. Nancy Reagan appeared. "Who are you?" she asked. "What are you doing in my backyard?"

Dana explained himself, assuring Mrs. Reagan that he and his friend would only need to talk to her husband for a couple of minutes.

"I know him," Mrs. Reagan replied. "He won't spend a couple of minutes with you. He'll spend fifteen or twenty. Then he'll either skip his breakfast or run late for his meetings all day." Mrs. Reagan promised to do what she could about Youth for Reagan herself, speaking to the campaign manager. Then she politely but firmly asked Dana and his friend to leave.

The two students picked up their sleeping bags, walked around to the front of the house, then began trudging down the long driveway to the street, dejected. Suddenly they heard footsteps. Turning, they saw Ronald Reagan pelting down the driveway to catch them. His shirttail was untucked. Shaving cream covered half his face. When he reached them Reagan said, "If you can spend the night on my back lawn, I can spend a few minutes talking with you. Now, what's the problem?"

Youth for Reagan was never disbanded.

The moral of Dana's story, I'd always thought, was simple. Whereas Nancy Reagan could prove hard and unattractive, Ronald Reagan always proved soft and winsome. But when I tried looking at the story not from Dana's point of view but from that of Reagan himself, the story got a little more complicated.

"How long did Reagan spend talking to the two of you in the driveway?" I asked Dana, going over the story with him once again.

"I don't know. Maybe fifteen or twenty minutes?"

"So Mrs. Reagan was right when she said he'd spend a lot of time with you?"

"Sure she was right. She said she knew her husband, and she did."

Nancy Reagan, I saw now, had created a situation in which her husband simply could not lose. If Reagan had let Dana and his friend disappear down the driveway, he would have been able to eat his breakfast and remain on time for his meetings. That would have been good. In running down the driveway after them instead, Reagan had been able to look like a hero to a couple of kids who would then tell the story to all their friends. That was even better. Nancy Reagan? She wasn't going to look very good either way. She was an intelligent woman. She must have known that. But she hadn't cared. If doing so helped her husband, she'd play the heavy.

The second story concerned a 1980 dinner party at the home of Justin Dart, a rich Californian who had backed Reagan for governor, then continued to advise him when he ran for President. The table was set for only five, the Darts themselves, Nancy Reagan, William Simon, the businessman who had served as secretary of the treasury for Presidents Nixon and Ford, and Arthur Laffer, the economist who told me the story. During dinner Laffer mentioned the failed 1978 New Jersey Senate campaign of Jeffrey Bell, who had served on the Reagan staff when Reagan ran for President in 1976. If Reagan had appeared in New Jersey on behalf of Bell, Laffer suggested, Reagan might have been able to turn Bell's campaign around. Interpreting the remark as a criticism of her husband, Nancy Reagan shot Laffer a look. Laffer attempted to change the subject. But Simon, who lived in New Jersey and had supported Bell, picked up where Laffer had left off.

"He went after Reagan," Laffer now says. "He said, 'Yeah, what the hell is this? What kind of loyalty is that to show to a former staffer? Reagan could have won the thing for Bell, but instead we ended up with that left-winger Bill Bradley [Bell defeated Clifford Case in the Republican primary but lost to Bradley in the general election].' "

Mrs. Reagan spent a few moments attempting to defend her husband politely, mentioning scheduling difficulties, how doubtful it had been in 1978 that Reagan had a following in New Jersey, and so on. Simon persisted. At length Mrs. Reagan dropped her napkin on the table, stood, and walked around the table to Simon. Then she bent to face him. "Don't you *ever* criticize my husband again," she said. "Do you understand me?" Stunned, Simon fell silent. Mrs. Reagan walked back around the table to her place, sat down, picked up her napkin, and continued her meal.

When I first heard that story, I looked at it from the point of view of the four people who'd found themselves feeling uncomfortable for the rest of the evening, the Darts, Arthur Laffer, and William Simon. Simon was a brilliant businessman, a philanthropist, a loyal supporter of the Republican Party, and a friend of Ronald Reagan. He was capable of an occasional tirade, but everybody knew that about him. Mrs. Reagan could have ignored Simon until he'd lost interest or changed the subject. Instead she'd created a scene.

When I tried looking at the story from the point of view of Ronald Reagan, who would have heard it later that night from his wife, Mrs. Reagan looked not impolite but admirable. Reagan was only months from taking office. His friends should have been rallying to him, not launching attacks on him for failing to make a single-handed attempt to save a sinking senate campaign. What must it have meant to Reagan to know that he could rely on his wife for loyalty so fierce and absolute?

"Look," Clark Judge said one day, making an observation I

thought good enough to record in my journal, "when you're President of the United States you've got to check up on everybody who walks through your door to make sure they're not trying to take advantage of you, you've got to keep an eye on your own staff to make sure they're always putting your interests first, and you've got to worry about a thousand details, from the menus at state dinners to the scene you want on the cover of next year's Christmas card. Reagan just isn't temperamentally suited to handle all that. His wife is. So she helps him and protects him. Who cares if she's hard on the people around him? I mean, really, from the vantage point of history, *who cares?* She takes care of him— and then he goes out and takes care of the Soviet Union."

—— *Frozen Wieners* ——

M Y PARENTS had always had a close relationship, and I see now that I learned a lot about marriage simply by observing theirs. But when I was working in the White House I thought of my parents' marriage as something that just *was.* Learn from my mother and father? The very idea. The Reagan's marriage, by contrast, fascinated me. And once I began to understand all that Nancy Reagan gave her husband—comfort, stability, companionship, loyalty—I began to wonder whether I could find someone who'd give the same to me. Self-absorbed as ever, you'll notice, I continued to think about *me.* But still, this was a start.

And my bachelor existence was beginning to pall. This startled me. Although vaguely aware that two or three years made a big difference in the life of a child, I'd never considered the possibility that two or three years would make a big difference in the life of a twentysomething. Yet by the time I was twenty-seven or twenty-eight, I'd begun finding it difficult to have much fun.

Josh Gilder and I had spent so much time with each other, my journals record, that we'd become capable of quarreling like old ladies. When we played squash, we'd argue about whether a shot had been out of bounds. When we'd go out to a movie, we'd dispute whether the screenplay had been any good. One summer we both found ourselves with lot of vacation time coming. Since neither of us wanted to vacation alone, we spent a couple of weeks together in France. Everywhere we went, I recorded in my journal, Josh wanted to go to fancy restaurants and buy expensive wine while I resisted, unwilling to spend the money.

"Food and wine," Josh would say testily, "that's the *point* of visiting France."

"Food and wine," I'd reply just as testily. "Once you've finished them, they're gone. What's the point in *that?*"

When my parents stayed in my apartment one Thanksgiving, my mother took me on a tour of the place as if it weren't my own home but a museum of single life. Dustballs in the corners of every room, in the backs of the closets, under the bed. Dust so thick on the windowsills that it looked as if they'd been upholstered in plush. Stains, deep stains, on every porcelain surface in the bathroom, and a wet, slippery grime coating the floor of the tub so thoroughly that when I used my thumbnail to scratch it off, revealing the surface of the tub itself, the effect was as striking as white paint on black velvet. Chastened, I spent half a day making my apartment look presentable. Yet even as I shoved the vacuum cleaner back into the closet, I knew that in a couple of months the apartment would look just as it had when my mother took me on her tour. During my first couple of years in

Washington, my journals indicate, my slovenly way of life never bothered me. But from then on it got on my nerves.

I recall one especially unnerving instance of bachelor existence to this day. Visiting a single friend one day after work, I reached his apartment ahead of him. Hungry, I decided to root around for something to eat—the code of bachelor life, of course, made this perfectly acceptable. Checking the refrigerator, I found nothing but beer and soda. Then I opened the freezer. Encrusted with frost and ice—my friend, I could see, had never heard of defrosting—the compartment was empty. Yet on the floor of the compartment I noticed a peculiar shape in the ice. I found a fork, then chipped away. Wieners—it was a half-empty package of Oscar Mayer wieners, buried under the frost and ice like the remains of a mastodon in the Siberian permafrost. Somewhere I'd read a magazine article stating that bachelors suffered more depression and illness, earned less money, and died younger than married men. This, I thought, was the life that awaited me, and I had a vision of myself as a potbellied old bachelor with bad breath, chipped teeth, fungus between my toes, and a half-empty package of wieners embedded in the ice in my freezer.

About this time a few of us speechwriters had a bull session in the hallway outside our offices. One of us had just rented the 1954 comedy *It Should Happen to You*, with Jack Lemmon, Judy Holliday, and Peter Lawford, and in talking about the movie we realized that there were a couple of similarities between the lives of Peter Lawford and Ronald Reagan. Like Reagan, Lawford was a handsome B actor who, although never a star of the first rank, worked steadily in motion pictures for a number of years, then saw his career dry up while he was still middle-aged. And like Reagan, Lawford, who was married to President Kennedy's sister Patricia, saw his first marriage end in divorce.

There, however, the similarities gave way to contrasts. While Reagan recovered from his midlife reverses to establish himself in two new careers, first television and then politics, Lawford accomplished little in the second half of his life, acting in forgettable television roles while spending a lot of his time drunk or high, a genial has-been. Reagan and Lawford differed in background and temperament, of course. But the comparison got us thinking. In *Where's the Rest of Me?*, Reagan himself describes the years following his divorce as the unhappiest of his life, writing about his "lonely inner world." Reagan, we saw, might easily have started on the same downward spiral in which Peter Lawford would later find himself trapped, becoming nothing but a genial has-been himself. What prevented him from doing so? Nancy Davis.

"For Reagan," Clark Judge said, "every experience of his early life confirmed his view of himself as a remarkably appealing and talented guy." Reagan succeeded in radio, became a movie star, and married a beautiful woman. And then his wife divorced him. "Every divorce is painful," Clark continued, "but the breakup of that first marriage turned Reagan's entire opinion of himself upside-down."

Once the protagonist of a success story, Ronald Reagan was suddenly a failure. By giving herself to him—by showing that she thought so highly of him that she was willing to devote herself to him for the rest of his life—Nancy Davis restored Reagan's sense of self-worth, giving him the confidence to rebuild his career. I'd already come to understand that Nancy Reagan served as her husband's indispensable companion and support. Now I saw that she had done still more for him than I'd imagined. In marrying Ronald Reagan, Nancy Davis had saved his life.

—— *At the Drinking Fountain* ——

WHEN I WROTE in my journal that I was afraid of becoming a potbellied old bachelor, I suppose I knew that I was being melodramatic. Yet living entirely for myself had proven lonely, unsanitary, and—I still couldn't get over this—boring. I knew now that I was missing out on something. Something big.

Instead of simply crossing the hallway to go, as it were, on road trips, I began asking a couple of the researchers out on dates (my English girlfriend, tired of a transatlantic romance, had long ago broken up with me). These dates entailed something at which I'd never become practiced or relaxed, namely, conversing with a woman other than my mother. Conscious of my awkwardness, I often tried to make the conversation less important than the surroundings, taking my dates to lavish restaurants; cheap when dining with Josh, I couldn't spend money fast enough when with a young woman. (Getting ready for one of these dates, I asked Josh for advice about the wine. He took it pretty well.) Talking to the researchers, I found, was a lot like using my high school French in France itself. The more I did it, the better I got. Eventually I was even able to stop seeing the other party as an alien.

In the middle of 1986, a new woman began working in the Appointments and Scheduling Office, where she helped sort through the hundreds of invitations the President received each week, deciding which would receive polite form letters of refusal—you'd be surprised how many young couples invited the President to attend their weddings—and which would be forwarded to the senior staff for serious consideration. I first saw her

173

at the drinking fountain—or rather, Josh first saw her at the drinking fountain. He and I were shooting the breeze when he suddenly broke off eye contact. Following his gaze, I saw the new woman just as Josh put his lips together to emit an admiring but silent whistle. She was brown-haired, brown-eyed, and beautiful.

Even though I was getting a lot of practice at speaking to women, somehow I couldn't find the courage to talk to the new woman, whose name, I learned when I asked around, was Edita. I'd been friends with the researchers before I started dating them, and I was sure they knew I was only going out with them for— well, for practice. (And since they all had plenty of admirers and boyfriends, I was also sure they were only going out with me to be nice.) Once I spoke to Edita, it would be the real thing. I'd be like an apprentice pilot going up solo for the very first time, and the mere thought made me freeze. While I was shopping for Christmas gifts one day that winter at the White House Historical Association, kitty-corner from the White House itself, Edita walked in. She was wearing a royal blue coat, her cheeks bright red from the cold. The encounter reminded me of watching Ronald and Nancy Reagan wave to each other from the Rose Garden, except that whereas the mere sight of Mrs. Reagan had made the President feel more alive, the mere sight of Edita made me feel like a zombie. As she paced the aisles looking for gifts, I shadowed Edita, remaining hidden. But even though she remained in the building for a good twenty minutes, I failed to work up the courage to speak to her. I also failed to remain hidden. When I finally asked her out the following spring—Josh, sick of hearing me talk about her, put a telephone in my hand, then announced that if I didn't call her then and there, he'd call her *for* me—Edita said she'd always wondered why I hadn't spoken to her at the White House Historical Association.

I don't know. We just fell for each other. I can't say we fell hard, because there wasn't anything hard about it. It was light, unforced, spontaneous, and fun—over and over again, my journals record, we'd find ourselves laughing. We'd get together for a cup of coffee most mornings in the Old Executive Office Building cafeteria, go for walks together over the lunch hour a couple of times a week, then go out on dates every evening we could. At first I gave Edita the same treatment I'd given my other dates, taking her to expensive restaurants so she'd feel the evening had proven worthwhile even if I was bad company myself. But we saw so much of each other that I started to run out of money. Sheepishly, one evening I suggested that we try someplace less expensive, like a Chinese restaurant. And do you know what she said to that? She said that sounded like even more fun.

Comfort, stability, companionship, loyalty—all the benefits that Mrs. Reagan gave to the President, I began to feel that Edita was giving to me. The odd thing was that I felt them all right away. I experienced a sense of companionship—of ease and pleasure in Edita's company—on our very first date, and soon after I realized that I somehow found it a comfort to have her near me. Stability? Before I began dating Edita, I'd discovered I was capable of feeling cranky and low for days at a time, experiencing a foretaste, I suppose, of the depression to which bachelors were supposed to be so much more subject than married men. But after I began dating Edita I found that my moods leveled right out. I felt much happier, of course, but also much more balanced and even. As for loyalty, Edita made a point of saying nice things about any speech she knew I'd written. I laughed every time, but secretly I felt pleased.

The benefit I recall most vividly of all came as a complete surprise: a sense of liberation from myself. For months my self-absorption had been closing in on me, making me weary of the one person whose company I could never escape. Edita gave me someone else to think about. I couldn't have felt any more exhilarated if she'd released me from a jail cell.

Mrs. Reagan, as we have seen, complemented her husband, attending to matters for which she was more suited than was he. Edita likewise took me in hand. I'd always been a little self-conscious about my taste in furniture and clothing, for example, and Edita very sweetly let me know that I'd been entirely justified in feeling that way. To furnish my living room, I'd gone to a Scandinavian design center and bought two baby-blue sofas and a big chair made of brown leather and chrome. Then, on the wall over one of the sofas, I'd hung the enormous bearskin that an uncle had given me. Edita considered the bearskin amusing. She considered the rest a provocation. She made a few telephone calls, selling the sofas and then, unable to find a buyer, giving the chair away. Our one argument concerned the light-blue seersucker jacket I liked enough to wear on our first date. It was weeks before I agreed to surrender the jacket to the Salvation Army, and when I finally did so I was only placating Edita, not admitting she was right. My friends, I felt certain, would ask what had happened to that jacket, proving they missed it. Nobody uttered a peep.

As we grew closer, Edita and I found ourselves talking about the President and Mrs. Reagan.

"What's amazing about the Reagans," I'd say after telling Edita about the encounter I'd witnessed in the Rose Garden, "is just how much they mean to each other. She's his whole world." Translation: I'm getting serious about this. Would you ever permit me to place you at the center of my life just as completely as the President places Mrs. Reagan at the center of his?

"What I find so admirable about Mrs. Reagan," Edita would say after listening to me tell Dana's story about camping out in the Reagans' backyard, "is that she always backs her husband up. She puts his interests first." Translation: I'm getting serious about this myself. And if you want me to be as loyal to you as Mrs. Reagan is to the President, buster, you'd better make sure you're just as deserving of loyalty.

Edita and I dated throughout my final year at the White House. During the following year, the first of my two years at business school, we visited each other so often that my travel agent started making cracks about having our bills for airfare framed and hanging them over his desk. Edita cut down our expenses during my second year by flying to California and remaining there, finding a job on the Stanford campus and an apartment nearby. And then, when I graduated, we consolidated our accounts, getting married.

After that events moved quickly.

One month after our wedding, we moved to New York, where I had a job waiting at the News Corporation.

Nine and a half months later, Edita gave birth to our first child.

And one month after that, I was laid off.

The position in which I found myself after losing my job in my early thirties bore a certain resemblance to the position in which Ronald Reagan found himself when his acting career sputtered to an end while he was in his early forties. Reagan was recently remarried. I was recently married. Reagan and his wife had just had their first child, a daughter. Edita and I had just had our first child, a daughter. Reagan relied on the patience, support, and forbearance of his wife for several years as he struggled to find his feet, performing walk-on parts in television programs and accepting a gig as the emcee of a Las Vegas floor show before finally landing the contract with General Electric that provided him with a stable base. I? I relied on the patience, forbearance, and support of my wife as I struggled to find my feet for a couple of years myself.

I set up a card table in the basement of my in-laws' house, starting work on a book about business school as I began searching for a new job, just as I've already explained. When I landed a position as a civil servant, becoming the public relations official at the Securities and Exchange Commission, I remained a scribbler in my off hours, continuing to work on what became my first book whenever I could. At the end of a year, two events transpired. I sold my book. This was good. And President George H. W. Bush lost his bid for reelection. This was bad, and I don't just mean because I'd voted for him. Once Bush left office, all his political appointees, of which I was one, would have to march out into the private sector to find themselves new jobs. I could hardly believe it. I mean, really. I'd been laid off *again.* And this time it hadn't been Rupert Murdoch who'd handed me my pink slip. It had been the American electorate.

Although Bush was defeated in November, he wouldn't leave office until January. That gave me a couple of months. I applied to advertising agencies, public relations firms, three or four big corporations, and a couple of investment banks, receiving half a dozen expressions of interest and a couple of firm offers, including an offer from an investment bank that would have paid very well. Then I heard from the director of the Hoover Institution, the think tank at Stanford. If I'd devote part of my time to helping him run the place, he said, he'd see to it that I'd be able to devote the rest of my time to writing. To Edita, the choice was never even close. During business school she'd seen me try investment banking, so she knew how much I disliked it, and during the past year she'd seen me devote as much time as I could to writing, so she knew how much I liked it. Without even mentioning the difference in income that doing so would entail, she insisted—*insisted*—I take the position at Hoover.

Nancy Reagan saved her husband's life. And my wife saved mine.

JOURNAL ENTRY, NOVEMBER 2002:

When I talked to Ron Reagan today, I told him how surprised I'd been when I visited Rancho del Cielo.

"I'd always assumed the ranch was close enough to allow your mother to drive down to L.A. whenever she wanted," I said. Ron laughed.

Instead, I'd found, the ranch is a good two and a half hours from Los Angeles—longer if you run into traffic—and the last seven miles lead up a badly paved mountain road with repeated switchbacks and hairpin turns. Drive back and forth for cocktails in Bel Air or dinner in Beverly Hills? Not a chance. When Nancy Reagan was at Rancho del Cielo, she was cut off.

"And I'd always assumed the ranch house was a really big, elegant place," I said. "Somehow it always photographed that way." Once again, Ron laughed.

Elegant? The floors are linoleum, the rooms smell musty, and the wall hangings include the mounted head of a longhorn steer, a polished slab of irregularly shaped wood fitted with hands and a dial, making it one of the ugliest clocks I'd ever seen, and a gun cabinet containing half a dozen rifles. My guide, Marilyn, explained that Mrs. Reagan had once retained Ted Graber, the interior designer, to spruce the place up. Graber hadn't gotten very far, adding only a couple of sets of red curtains and a few pieces of furniture upholstered in fabric with a western motif.

The most striking sight? The master bed. It was just two old twin beds that Ronald Reagan had shoved next to each other. To hold the beds together, Marilyn pointed out, Reagan had fastened the cast-iron head-boards together with plastic ties of the kind you might use to bundle electrical cords.

"Ron," I said, "I kept thinking, 'Nancy Reagan? In this *place?' "*

"And she had to put up with that sort of stuff for their entire marriage," Ron replied.

"The ranch house in Malibu {where Reagan owned a ranch before purchasing Rancho del Cielo} was our childhood ranch house—the place where my sisters, my brother, and I grew up," Ron continued. "It was the kind of place where the foundation was collapsing and there was a hive of bees in the closet and dead bees all over the floor in the kids' room. In retrospect, it was just this hovel.

"Those ranches are one measure of my mother's love for my father."

A S I DROVE north on Highway 101 to return home from the Reagan ranch last May, I found myself considering all that Ronald Reagan had taught me about marriage. Reagan had shown me what marriage can do for a man, enabling me to grasp that the self-absorbed existence of a bachelor might not have been the highest good after all. Then, after business school, when I found myself laid off twice in two years, I'd learned from his example once again. At first I'd resented having to rely on my wife. I'd always appreciated her patience and support, but now I'd *needed* them. That had changed things, placing me in a position that had seemed—well, that had seemed unmanly. But Reagan was a man's man. If he'd relied on his wife, I'd decided, then I ought to be able to rely on mine. And now, at Rancho del Cielo, I'd learned from Reagan's example yet again.

You see, I'd always imagined that by this stage of my life I'd have given my wife a big, beautiful house—if I'd pictured myself as Fred MacMurray, I'd pictured Edita in a bungalow like the one on *My Three Sons*. We do own a house. But we bought it after the high-tech bubble had inflated housing prices in Northern California to such high levels that Palo Alto, the town where we live, became one of the most expensive communities in the country. Forced to make our purchase when we did—we'd had a fourth child, outgrowing the house we'd been renting—we'd pulled together all our savings, then shopped around for the biggest mortgage we could find. Even at that we'd only been able to afford a forty-five-year-old ranch house that needed to be completely remodeled. *My Three Sons?* Try *Green Acres*. And that isn't the worst of it. After spending so much just to get into the place, we're going to have to do a lot of the remodeling ourselves. Carpenters? Painters? Electricians? Just my wife, me, and a how-to book from Home Depot.

Edita claims she loves the fixer-upper, even the idea of fixing it up ourselves, and while I'd gone out to collect interviews for this book she'd gone out to collect samples of paint and fabric for the house. I'd assumed she was just making the best of a bad situation. After seeing the Reagan ranch, though, I wasn't so sure.

When Reagan purchased Rancho del Cielo in 1973, he was a wealthy man. He could have had the tiny old adobe ranch house torn down, then had a big, elegant ranch house, designed, conceivably, by Ted Graber, erected in its place. Reagan had chosen to keep the tiny old adobe ranch house all the same. He'd hired a contractor to enclose the porch and enlarge the master bedroom. But with the help of only Barney Barnett and Dennis LeBlanc, his two longtime hired hands, Reagan himself had ripped out the porch, replaced the roof, put down the linoleum floor, painted each of the rooms, and laid the new flagstone patio, devoting to the project one weekend after another. Three men, rebuilding the place on their own. At least that's the way I first heard the story. Then, when I talked to Dennis LeBlanc, I learned that the men had usually had their meals with Mrs. Reagan. "We'd have lunch out on the patio and dinner at the table in that other room [that was formed by walling in the porch]," LeBlanc explained, "and if there was something we'd want to see on television, then we'd have our dinner on trays in the inner room where the television was. Once we got the ranch house to where it could be lived in, Mrs. Reagan was there most of the time."

Remodeling the ranch house hadn't been an exclusively male effort after all: Reagan had shared the project with his wife. Talking with Mrs. Reagan about the color each room should be painted, asking her where she'd like him to place the new patio or how she wanted him to arrange the furniture—Reagan, I saw, had made remodeling the ranch house into a joint undertaking, less concerned with the way the place looked than in making it *theirs*. And even though his wife's tastes ran to comfort and elegance, not ruggedness and rusticity, she had loved the place.

"The last time she was up there [after the ranch had been sold]," says John Barletta, a Secret Service agent assigned for many years to Rancho del Cielo, "I said to my people in the Secret Service, 'When she leaves, she's going to cry in the limo. Just have some water there for her. Just let her go and make sure she's all right.' And sure enough, they called me the next day and said she cried practically the whole way home."

Edita, I realized, really did love our fixer-upper. I could stop wishing I'd been able to give her a big, beautiful house and instead join my wife in making our house *ours*.

T WENTY YEARS AGO, I thought as I drove, I didn't have a care in the world. Now I had to support a wife and five children—since moving into our house, Edita had had our fifth—and when I finished this book I'd have to repaint our house inside and out, repave the driveway, put in a new lawn, and find somebody who could replace our windows.

And, I realized, I'd never been happier.

—————— *Eight* ——————

THE OAK-WALLED
CATHEDRAL

Say Your Prayers

JOURNAL ENTRY, MAY 2002:

My conversation with Josh Gilder a couple of weeks ago proved typical. I asked Josh if he believed Reagan was a man of faith.

"I'm sure he was," Josh said. "It was obvious."

"But why?" I replied. "Why was it obvious?"

There was a long silence.

"You know," Josh finally answered, "that's a really good question. We all sensed it—we just knew. But it's not as if Reagan wore his religion on his sleeve, is it?"

Josh Gilder, Clark Judge, Dana Rohrabacher, Bill Buckley, Ed Meese, Lyn Nofziger. Each considered Reagan a man of faith, but none could say just why. Today I talked with someone who could.

——*In the Cathedral*——

WILLIAM CLARK met the future President in 1965. When the then-candidate for governor asked Clark to serve as Ventura County chairman of his campaign, Clark agreed to join him for a trail ride in the Ventura hills to discuss the matter. During that trail ride, something happened to Clark that happened to very few. He became close to Ronald Reagan.

"He reminded me so much of my own father," Clark said when I visited him at his ranch near Paso Robles. A fourth-generation Californian, Clark had helped his father work the horses on the family spread. "I'd be in the corral. My father would open the gate to let a horse in, and somehow just from the look on my father's face I'd know how he'd want me to work that horse. That's the way it was with Ronald Reagan. Not a lot of words. Just glances and gestures. Somehow we just seemed to read each other."

Clark served Governor Reagan as cabinet secretary and chief of staff, leaving the governor's office when Reagan made him a judge (Reagan named Clark to the Superior Court of San Luis Obispo County, then to the California Court of Appeals, and finally to the California Supreme Court). Clark then served President Reagan as deputy secretary of state, national security adviser, and secretary of the interior. Even after returning to private life at the beginning of Reagan's second term, Clark remained so close to Reagan that he often operated as a personal emissary or envoy for the President. (One afternoon in 1987, more than two years after Clark left Washington to return to California, one of us speechwriters got back from lunch to find that a confused White House operator had sent him nearly a dozen telephone messages intended not for him, the speechwriter, Clark Judge, but for the former national security adviser, Judge Clark. The callers had included the Queen of England, Prime Minister Thatcher, and President Mitterand.) "Reagan came to look on Bill Clark pretty much as a son," Lyn Nofziger says. Roger Robinson, on the staff of the National Security Council during Clark's tenure as national security adviser, says the relationship between Reagan and Clark prompted him to think back to the Eastern religions he had studied in college. "That was the only thing that helped me to understand those two men," Robinson says. "Reagan and Clark were so close it was zenlike."

As Judge Clark and I talked, I found myself wanting to ask not about Reagan's policies but about his interior life. What had Clark, the man who was probably closer to the President than anyone outside the Reagan family, seen in the chief executive that would have been hidden from an ordinary member of the staff such as me? The private, inner Reagan—what had he been like?

"He was a man of prayer," Clark said.

Reagan, Clark explained, prayed in all moods and in all circumstances. Even during government meetings, Reagan might offer a brief, silent prayer. "I could usually tell when he was in communication with our Lord," Clark said, chuckling. "When he was leaning back his head and looking at the ceiling, that's often when he was praying."

Flying with Reagan from Sacramento to Washington for a governors' meeting in 1968, Clark was startled to see the captain emerge from the cockpit, then walk down the aisle to find him. "He said, 'Mr. Clark, you might want to tell this to the governor. We've just received word that Martin Luther King Jr. has been

186

shot and killed.' " Clark stepped back to inform Reagan, who was seated in a row behind him. "I expected him to say something, but he was silent," Clark said. "He just looked down at his feet." Clark returned to his seat. When he glanced back a moment later, Reagan's head was bowed, his lips moving. "He was in prayer."

When in 1982 Leonid Brezhnev died, Clark joined the President and Mrs. Reagan in walking the couple of blocks up Sixteenth Street from the White House to the grim old mansion that served as the Soviet embassy. Once each of them had signed the book of condolences, the President made a suggestion. "He looked over at me with that wonderful catbird grin of his," Clark said, "and he asked, 'Do you think they'd mind if we just said a little prayer for the man?' " The President, Mrs. Reagan, and Clark bowed their heads. Then Reagan asked the blessings of God on the deceased leader of a nation devoted to atheism.

Reagan's favorite setting for prayer? The outdoors. "He didn't need a church to pray in," Clark explained. "He referred to his ranch as an open cathedral with oak trees for walls." On trail rides, Clark and Reagan would often recite the famous prayer of St. Francis of Assisi that opens, "Lord, make me an instrument of Thy peace." "Sometimes," Clark said, "the President would look around and say, 'What a wonderful place for prayer.' And sometimes he'd just look up at the sky and say, 'Glory to God.' "

—— My Life as an Atheist ——

BOTH MY PARENTS believed in God. They read their Bibles, said their prayers—my mother even kept a list of the people for whom she wanted to be sure to pray each morning—and took me to church on Sundays. I never felt much urge to rebel. The existence of God seemed obvious. When I said my own prayers, I always felt someone was listening. When I went for a hike in the woods or watched the sun come up—our house stood on top of a hill, and my father and I used to like watching the sunrise as we ate breakfast—I always felt someone was in one way or another responsible for nature. In Vestal, New York, there was nothing remarkable about any of this. All the families in our neighborhood went to church, except for the Rosenthals, who went to synagogue, and all my friends at school believed in God—at Vestal Central High School, it would no more have occurred to us to doubt the existence of a Creator than to doubt the existence of the United States.

When I went away to college, I made a couple of discoveries. Strictly speaking, the discoveries had nothing to do with each other. To this day, though, I can't think of one without recalling the other. The first was that the cool kids, the kids who had gone to prep schools such as Andover and Exeter, all wore trousers of a kind I'd never noticed before: khakis. Back home in Vestal, we'd all worn jeans to school. If we'd wanted to appear dressy—if, for example, one of us, running for student council, had to make a speech—we'd worn corduroys. Khakis? I wasn't even sure how to pronounce the word. The other discovery was that a lot of my professors didn't believe in God. Just how I first became aware of this, I can no longer recall. Maybe it had something to do with the way so many of them talked about existentialism. Existentialism, I learned, held that life was absurd. Like Sisyphus, the cruel king condemned in Hades to roll a rock up a mountain every day, only to see the rock roll right back down every night (I took a class in which Albert Camus's *The Myth of Sisyphus* was on the reading list) each of us was doomed to a senseless existence, our lives devoid of any meaning other than the meaning we ourselves imputed to them. Many of my professors, I found, considered this view sophisticated and brave.

How did I respond to these discoveries? By remaining true to myself, sticking to jeans and going to church? No. I bought two pairs of khakis and became an atheist.

Neither the khakis nor the atheism worked out. Informal attire though anyone else would have considered them, khakis made me feel as if I were putting on airs. The atheism? I tried— I really tried. But God kept coming to mind. Every time He did, I had to stop and think, reminding myself that He didn't exist. The turning leaves, the sharp breath of the coming winter, the fascination of making new friends, the glorious spectacle of an Ivy League football game—all the pleasures I was encountering as a Dartmouth freshman, I had to tell myself again and again, had come about by accident, random products of the whirling of atoms. The more I tried to avoid thinking about God, the more I found Him on my mind. An atheist, I became as preoccupied with God as if I'd been a Jesus freak. My studies suffered. I found it harder and harder to have any fun. At the end of exactly two weeks, I said the hell with it and went back to church.

Even though I was never able to get rid of my faith, however, I did learn to feel embarrassed about it. If the subject of religion came up when I was with people I regarded as especially intelligent or cool, I'd keep my mouth shut. Under direct questioning, of course, I'd own up, but I was no happier to admit I was a believer than to admit I'd come from a small town. Faith was unsophisticated. I didn't want to be unsophisticated.

This brings me to Ronald Reagan.

— *Wormwood, the Pope,* — *and the Chief of Staff*

THE FORTIETH CHIEF executive may not have worn his religion on his sleeve, but you couldn't write speeches for him without seeing over and over again just how fundamental it was to his outlook. Consider three incidents that took place during speech meetings.

One occurred soon after the 1986 nuclear accident at Chernobyl. "I just learned that *'chernobyl'* is the Russian word for 'wormwood,'" the President said, rising from his desk as we speechwriters filed into the Oval Office. He walked to his chair near the fireplace. "'Wormwood' is mentioned in the Book of Revelation, you know," he said, sitting down. "It's one of the seven plagues that signals the end of the world."

About to take a seat myself, I froze for a moment. I wasn't exactly suspended between heaven and earth, just between an upright position and a big white sofa cushion, but I had a momentary out-of-body experience even so. We speechwriters all looked at each other. None of us said anything. What could we have said? I mean, what words were supposed to come to mind when you walked into a meeting with the President of the United States during the Cold War only to hear him begin musing about the end of the world?

The President looked from face to face, his eyes twinkling. "Kind of interesting, isn't it?" Then he nodded to the director of communications, signaling that the meeting could begin.

Asking around after this incident, Josh Gilder and I learned that all the old Reagan hands had heard the President mention

190

the Book of Revelation, the apocalyptic book with which the New Testament ends, often enough to suggest he had a particular fascination with it. (Working on this book, I learned from John Barletta, one of the Secret Service agents assigned to Rancho del Cielo, that Reagan even talked about the Book of Revelation on trail rides.) Did Reagan believe the Book of Revelation foretold present-day events? That St. John, the author, had predicted the disaster at Chernobyl nineteen centuries before it took place?

"Of *course* not," Josh said. "If he were a literalist about the Bible we'd have known it by now."

Neither Josh nor T could recall a single instance in which Reagan had ever based a decision on a biblical prophecy. According to many of those who believed in a literal translation of the Book of Revelation, for instance, the big, raspberry-colored birthmark on Gorbachev's forehead was the "mark" foretold in Chapter 13. Yet Reagan had chosen to befriend the man, not exactly the approach he'd have taken if he considered Gorbachev an instrument of the devil.

"Reagan doesn't keep going back to the Book of Revelation because it's some sort of mechanically accurate prophecy," Josh continued. "He keeps going back to it because it shows that somehow or other human history is in God's hands." Scripture contained hints about the future, but only hints. Yet scripture unambiguously portrayed history as a divine drama in which, over the course of the centuries, the will of God would prevail. "Reagan already knows how he intends to play his part in the drama," Josh said. "And he has the curiosity of a fellow actor about how God intends to play His."

Josh himself prompted the second incident. Preparing remarks for the President to deliver during his 1987 visit to the Vatican, Josh had done all he could to rework the bland material the State Department had given him, but he remained unhappy with the results. "Everything was 'Hello, it's great to be here in the Vatican, how's everybody doing?' " Josh recalls. "The remarks didn't *say* anything."

During the next speech meeting Josh piped up, asking the President what role he believed religion might play in the reform of Eastern Europe. "I was writing really fast," Josh says, "and I managed to get down just about everything he said." (This incident took place during the same meeting in which my draft of the Berlin Wall address was discussed. You'll recall how miffed I felt that Josh got the President talking when my own attempt to do so failed.) Back in his office, Josh placed the President's comments in the middle of his draft, transcribing them almost word for word.

> Our prayers will go with you [the Pope would soon be traveling to Poland] in the profound hope that soon the hand of God will lighten the terrible burden of brave people everywhere who yearn for freedom, even as all men and women yearn for the freedom that God gave us all when he gave us a free will. We see the power of the spiritual force in that troubled land [of Poland], uniting a people in hope, just as we see the powerful stirrings to the East of a belief that will not die despite generations of oppression. Perhaps it's not too much to hope that true change will come to all countries that now deny or hinder the freedom to worship God. And perhaps we'll see that change come through the reemergence of faith, through the irresistible power of a religious renewal.

" 'The freedom that God gave us all when he gave us a free will.' Beautiful, right?" Josh says. "I mean, it was right up there with the Declaration of Independence and 'All men . . . are endowed by their Creator with certain unalienable rights.' "

The incident involved a postscript too rich to omit. When Josh sent the remarks out for staffing, the State Department cut them almost in half. "The whole thing about spiritual renewal had been X'ed out," Josh says. Incredulous, Josh telephoned the responsible State Department official. Josh retains a vivid memory of the conversation.

"I just have to know," Josh said. "Why did you take out the material about spiritual renewal?"

"I think it's inappropriate to have so much language about God," the official replied.

"So much language about God?" Josh said. "The President will be talking to the *Pope.*"

Only when Josh informed him that the material had been dictated by the President himself did the State Department official relent.

The third incident took place in 1984, before either of the other two, but I'm placing it in the final position because I remember it so vividly. I would. It concerned a speech I wrote.

Before composing the speech, I met a couple of representatives of the organization before which the President would be appearing, the National Association of Evangelicals. Were there any issues in particular, I asked, that their members would want to hear the President address? They replied that everyone in the National Association of Evangelicals so loved Ronald Reagan that the President could speak about defense, foreign policy, economics, or any other topic. There was only one caveat. The President would need to include at least a few sentences about his opposition to abortion. If he didn't, the audience would feel let down.

I drafted what amounted to a campaign speech—this was

1984, as I've said, the year Reagan was running for reelection—
that reviewed the President's record, discussed what he hoped to
accomplish in a second term, and presented a statement of his
most fundamental convictions. Under this last heading, I worked
in a mention of abortion. "We must do our duty to generations
yet unborn," I wrote. "Abortion as a means of birth control must
stop."

Seated during our next speech meeting in his usual place, the
chair to the left of the fireplace, the President appeared relaxed,
jovial, gentle, serene. Commenting in an unhurried manner on
this or that speech as the meeting progressed—the meeting covered half a dozen speeches—he looked as though he'd have been
happy to sit there and talk about speeches all day. Seated on a sofa,
Dick Darman, then director of communications, looked by contrast tense, even for Darman, which was saying something, while
chief of staff James Baker, seated in the chair to the right of the
fireplace, next to the President himself, looked as if he were having as much trouble keeping still as a hyperactive child. Baker
fidgeted, glanced at his watch, crossed his legs, uncrossed his legs,
then glanced at his watch yet again. Busy though they may have
been, I found myself thinking, Darman and Baker looked almost
too anxious. And why was Baker even present? This was the first
speech meeting I'd known him to attend. Darman would introduce each speech, the President would offer his comments, the
speechwriter concerned would ask a question or two, and Baker
would say . . . nothing. Glancing at his watch once again midway
through the meeting, Baker stood, then, without excusing himself, walked from the room. Two or three minutes later he came
back—I supposed he'd placed a couple of telephone calls—
returned to his chair, and, once again, just sat there.

When we'd reached the speech just before mine, Baker glanced at his watch, then began standing as if he intended to leave the room once again. "Not now, Jim," Darman said under his breath—I could hear only because I was seated across from him—"Robinson's speech is next." Robinson's speech? The chief of staff had attended this meeting because of Robinson's speech?

The discussion of my speech proved routine—the President said he liked it, then asked, as I recall, for more emphasis on the economy. Just as I was about to decide I'd misheard Darman's whispered remark to Baker, the director of communications turned to the chief of staff.

"Jim," Darman said, "I believe you have a couple of thoughts on this one?"

"I do," Baker replied. And then he launched into a little speech of his own. It was so smooth that I wouldn't have been surprised to learn he'd practiced it in front of a mirror.

The day after the President was to deliver the speech I'd drafted, Baker said, Congress would vote on the school prayer amendment, a measure intended to make prayer in public schools constitutional. This accident of timing would enable the President to discuss the school prayer amendment in his speech. And just as soon as the President had concluded his remarks, everyone in the National Association of Evangelicals would stampede to the telephones, calling the church folks back home. In a matter of hours, Congress would find itself overwhelmed with telephone calls, faxes, and telegrams in support of the school prayer amendment.

While Baker spoke, we speechwriters darted looks at each other in disbelief. Baker? Urging the President to speak on behalf of the school prayer amendment? Baker was a pragmatist, not a true believer. He spent his time trying to moderate the President's conservatism, not shore it up. Reagan listened to Baker serenely, of course. But even he looked puzzled.

"And if you talk about the school prayer amendment," Baker said to the President, drawing his little speech to a close, "then you won't have to mention abortion." *That* was it. Baker didn't care about the school prayer amendment. He only wanted to talk the President out of raising a divisive issue.

After this crude attempt to manipulate him, the thought crossed my mind, the President might engage in one of his rare displays of anger. Yet Reagan appeared amused, for all the world as if a teenager had just tried to bamboozle him into handing over the car keys. "Now, Jim," the President said, shaking his head and smiling, "this is just one of those things I feel very strongly about—"

"But, Mr. President," Baker said, cutting off the chief executive, "all the polls show the country just isn't with you on this one." Baker launched into another smooth little speech. A majority of women disagreed with the President on abortion, Baker said—among single women, the percentage was higher still. And now that the reelection campaign would soon be getting under way, the President should avoid antagonizing any group, particularly one as large as the female population. The President's last statement about abortion had caused enough trouble as it was.

The trouble Baker mentioned had arisen when the President delivered a speech to the National Association of Religious Broadcasters just over a month before. Composing the remarks, Bently Elliott, then director of presidential speechwriting, had included a couple of paragraphs about new evidence that fetuses could experience pain much sooner than had previously been supposed. We must recognize, Ben's draft argued, "the excruciating pain the unborn must feel as their lives are snuffed away." After the President delivered this speech, every abortion rights group in the country had denounced him.

"The country already knows where you stand on abortion, Mr. President," Baker said. "There's just no reason to go into it all over again."

"Well, Jim," the President said, shaking his head and smiling once again, "I just don't know about that." All human life was sacred, the President explained, and until somebody could prove to him that fetuses weren't human, he'd continue to oppose abortion. "Come to think of it," he said, "I got a letter just the other day you might find interesting."

The President stood—the rest of us began to stand, too, just as protocol required, but the President waved us back to our seats—then crossed the room to leaf through some papers on his desk. "Here it is," he said, lifting a sheet of stationery. He read the letter aloud. Despite the uproar his recent speech had provoked, the letter stated, the President had been correct to assert that fetuses could experience pain much sooner than had previously been supposed. Signed by more than half a dozen medical doctors, the letter concluded by thanking the President for having the courage to speak out. "Why don't we mention this letter in that speech we've just been talking about?" the President asked. Crossing the room to return to his chair, he handed the letter to Darman.

The President had never so much as raised his voice. Yet in the full knowledge that he would pay a political price for doing so, he had overruled a recommendation from his own chief of staff. Human life was sacred, and that was that. James Baker registered no emotion, but his face seemed to tighten, as if someone had twisted a knob on the back of his head, increasing the tension on his scalp. Less adept at concealing his feelings, Dick Darman blanched. Ronald Reagan? He appeared just as serene at the end of the meeting as he had at the beginning.

This incident, too, involved a postscript too rich to omit. Still furious the following morning, Darman summoned Ben Elliott and me to his office, then berated us for a quarter of an hour, threatening to fire us. (Maybe he did fire us. I thought he did, anyway. "For about thirty seconds there," I asked Ben afterward, "weren't you and I unemployed?") Somehow I found the courage to ask Darman about an aspect of the incident that had puzzled me. If he had objected to my passage about abortion, why hadn't he simply told me to take it out before he sent the speech to the President? The question only made Darman angrier, forcing him to admit that, although meticulous about paperwork, for once he had permitted a document to reach the President without examining it himself. "Don't you understand?" Darman railed. "It's our *job* to protect the President from himself." That sentence is a direct quotation.

Musings about wormwood, a message for the Pope, and an avuncular but pointed rebuke of the chief of staff. Taken together, the three incidents amounted to a kind of catechism. Reagan could wonder aloud whether the disaster at Chernobyl was somehow hinted at or foreshadowed in the Book of Revelation because he believed so implicitly in the power of God. We might not always be able to understand quite what He's up to, but history lies in His hands. The President's message for the Pope? We possess our rights, Reagan believed, not by virtue of some government diktat, but by virtue of our innate dignity as children of God. The chief executive's rebuke of James Baker may have proven the most telling incident of the three. It demonstrated Reagan's conviction that human rights include the right to life. But it also demonstrated something basic about Reagan himself. Despite his respect for the will of the American people, he held himself accountable to his own conscience—and, ultimately, to God.

Ronald Reagan may never have invited his secretary of state, George Shultz, to join him in prayer the way Richard Nixon once invited Henry Kissinger to join him in dropping to his knees. He may never have refused to serve liquor in the White House, like Jimmy Carter, or spoken in public about his need for repentance, like Bill Clinton. Yet if Reagan never wore his religion on his sleeve, he never gave the slightest indication that he ever felt embarrassed about it, either. His love of God proved as central to the way he looked at the world—as central to his very being—as did his love of country. Working for him, I was finally able to quit feeling embarrassed about my own faith. Whenever I'd hear someone speak as if only rubes believed in God, I'd remind myself that Reagan had succeeded in radio, motion pictures, television, and politics—and that it looked as though he was winning the Cold War. That was a lot more than Albert Camus had ever accomplished. I'd never read Camus in French, so he might have been a greater writer than I recognized. But you see what I was saying.

——*Why I'm Not a Beach Bum*——

WORKING FOR REAGAN taught me a second lesson about faith, one I still think about almost every day. It enabled me to see how faith and everyday life fit together.

I used to spend a lot of time worrying about this. The problem always took the same form, presenting itself as a question. It was a simple question, but you could spend hours trying to answer it—at least you could if you had a turn of mind like mine. If God were omnipotent, the question went—if He controlled every aspect of human history, from the first moment of creation until the very crack of doom—then why not just sit back, relax, and let Him work everything out? Why even bother showing up at the office in the morning? Why not move to southern California and become a beach bum instead? No matter what we did, the future would still unfold just the way God intended, wouldn't it? I had no intention of becoming a beach bum, of course. But it bothered me that I couldn't see why I shouldn't. If only I had enough faith—if only I *really* believed—wouldn't I quit my job and buy a surfboard?

What I'd done was stumble, all on my own, onto what I eventually learned was one of the hardest of all theological problems, the opposition or contradiction between predestination and free will. The doctrine of predestination holds that the future is foreordained; the doctrine of free will that men can make fundamental choices about their lives and that the future, in turn, is therefore open-ended and unknowable. How can both be true? Some of the greatest minds in history, including St. Augustine, Maimonides, and St. Thomas Aquinas, had all gotten hung up on this problem. Knowing that didn't help. I went right ahead and got hung up on it myself. When the doctors of the Church called predestination and free will mysteries, I saw, they weren't kidding.

Now, Ronald Reagan never solved the problem for me in the abstract—as far as that goes, I never came across any evidence that he gave the problem any thought. What Reagan did instead was show me how to handle the problem in practice. Predestination? Free will? He simply believed in both. Let me tell you, for example, what I witnessed one late October afternoon in 1987.

Aboard Air Force One on a day-long trip from Washington, D.C. to the United States Military Academy at West Point, I found myself seated next to a woman in her early sixties. For several years, she explained just after takeoff, she had served as a volunteer in the correspondence unit, helping to open, read, and sort the thousands of letters that arrived at the White House each week. Now she was leaving the job, and, as a gesture of gratitude for all the work she'd done, her supervisor had arranged for her to accompany the President on this trip. "I'm so excited," she said. "I've never met the President. But everyone told me he always comes back to greet people."

When Air Force One was somewhere over northern New Jersey, the President did just that, entering the rear cabin. He was as immaculately dressed and groomed as ever. Even so, he looked bad. His face appeared haggard, his eyes red. Less than two weeks earlier, Mrs. Reagan, who had been diagnosed with breast cancer, had undergone a mastectomy. Then, just nine days after the surgery, Mrs. Reagan's mother, Edith Luckett Davis, had died. In the space of two weeks the President had had to help his wife recover from a major operation and then support her as she grieved the loss of a parent. Ordinarily Reagan looked ten years younger than his age. Now he looked ten years older.

When he reached me, the President shook my hand. This produced an awkward moment. He seemed too weary to make conversation; I felt too shocked by his appearance to know what to say. Then he reached across me to shake hands with the woman from the correspondence unit. She responded to his appearance not with a loss of composure but with a spontaneous and unfeigned display of concern. She introduced herself, told the President that everyone who worked with her was sorry to have heard about the death of Mrs. Davis, and then asked simply, "How is Mrs. Reagan holding up?"

Touched, the President found the energy to spend a few moments chatting. Although Mrs. Reagan was still in some pain, he said, the doctors were pleased with her progress. Yet the loss of her mother had hit her hard. He had told her, he explained, that God's plan for each of us included the moment of our passing. "When the Lord closes the door on this life," the President said, "He just opens the door on another and leads us right through." Then he smiled warmly, thanking the woman for her concern, and continued down the aisle.

When Air Force One landed, the President looked himself again. He stood at the door of the aircraft to wave for the cameras, jauntily descended the steps, then snapped off a salute to the honor guard waiting for him on the tarmac. When he reviewed the cadets as they marched past him on the parade ground, he projected energy and vigor. And then, when he delivered the speech I'd drafted, he did so with conviction, describing progress in talks with the Soviets (just a few weeks later the President and Gorbachev would sign the INF treaty), the importance of maintaining our military might, and the challenges that would fall to the next generation of officers. "Your youth, your optimism— they give me strength" he said. "As I look out upon your young faces, I feel as one who will depart the stage almost before you've made your first entrance. I feel in my heart a great confidence in the future of our country, for I know that you will defend that future."

Predestination? As he told the woman seated next to me on Air Force One, Reagan believed that God has a plan for every person. Our lives, even our lives after death, lie in His hands.

Free will? As his comportment demonstrated—although tired and sad, he made certain he projected vitality—the President believed that the success of the afternoon depended on him. And as the conviction with which he delivered his speech demonstrated, he believed that peace and freedom throughout the world depended on the United States, and that the future of the United States, in turn, depended on the cadets arrayed before him. God may have been in charge, but somehow, Reagan held, it was all still up to us.

Not long after that trip to West Point, I noticed a card that one of the secretaries in the office had tacked to a bulletin board over her desk. "Pray as if everything depended on God," the card said. "Work as if everything depended on you." That, I thought, was a neat summary of the way Ronald Reagan led his life, and ever since I've recognized that habit or pattern of life as an ideal. I don't pray as much as I should—for that matter, I don't work as much as I should. But as erratic or fitful as my efforts always prove, I try.

JOURNAL ENTRY, MAY 2001:

Ever since my talk with Judge Clark, I've found, a picture keeps coming to mind. Ronald Reagan is on horseback, riding along the exposed ridge at the southwestern corner of Rancho del Cielo. When he reaches the high point where the helicopter pad once stood, he reins in his mount. He gazes up at the enormous vault of the sky. He feels the rushing wind against his face. He looks east, following the shape of the land as it tumbles down and away, spreading to form the green bowl of the Santa Ynez Valley. Then he shifts in his saddle to look west, taking in the endless, dazzling ocean, the Channel Islands misty in the distance. And then he whispers: "Glory to God."

———*Nine*———

TOMFOOLS

*Do What You Can, Where You Are,
with What You Have*

JOURNAL ENTRY, MARCH 1983:

*As we waited outside the Oval Office for a speech meeting, a certain mem-
ber of the senior staff* [whom I'll call "the pragmatist"] *looked over the
outlines we'd be discussing with the President.*

*" 'Unalienable rights?' " he asked, pointing to a phrase in my out-
line from the Declaration of Independence. "That should be 'inalienable
rights.' When you prepare material for the President of the United States,
Peter, check your spelling."*

*"But look closely," I replied. "I placed the phrase in quotation marks.
Whether it was standard eighteenth-century spelling or a printer's mistake, I don't know. But the word in the Declaration is 'unalienable.' "*

*The pragmatist's face fell. He turned his back on me. And ever since
this incident—which took place a good three months ago now—he has refused to speak to me, even to say hello in the hallway.*

"It was just a two-word quotation," I told one of his friends today. "That couldn't be why he won't speak to me, could it? Something as petty as that?"

She smiled knowingly. "In his mind," she said, "he's never left the Harvard faculty. He hates to be wrong about anything."

Petty, peevish, small-minded, arrogant. As far as I'm concerned, the list of adjectives that applies to the pragmatist reads like the entry in Roget's Thesaurus *under the heading for "jerk."*

—— *Splat, Thud* ——

HUMAN IMPERFECTION. When I started working at the White House, I wasn't sure what to make of it. I'm not talking here about depraved or criminal behavior. I'm talking about petty, annoying, ordinary human imperfection of the kind you come across again and again in everyday life. When I encountered it, I just didn't know how to respond.

My first year at college, I lived directly above a Peter Frampton fan named Gary. Every afternoon when he'd get back to his room after classes, Gary would put on the same Peter Frampton album, drop the needle of his stereo system into the first groove of the same song, and crank up the volume just as high as it would go. I can't tell you any of the lyrics. They never quite penetrated the floor. But the opening notes of the base guitar went like this: "Da da da DA duh duh, da da da DA duh duh, boom, BOOM, BOOM."

I'd be seated at my desk, studying intently in one of my Type A phases—come to think of it, it was Gary, a pre-med student, who told me I had a Type A personality in the first place—when Peter Frampton would begin blaring. My room would throb. My windows would vibrate. I'd stamp my foot, trying to get Gary to tune down the volume. Gary would ignore me. So I'd climb onto my chair, jump, and land with a thud. Gary would emit a howl— as he had made the mistake of telling one of my friends, every time I jumped on my floor sawdust fell from his ceiling, covering his room in a coarse dust like volcanic ash. Sometimes Gary would relent, tuning Frampton down. But sometimes he'd dart into his bathroom, soak a roll of toilet paper under the faucet, then step outside to begin whipping wads of toilet paper at my windows. Each time a wad would strike, adhering to the glass with a splat, I'd climb onto my chair and jump. *Splat, thud. Splat, thud.* Even when Gary finally tuned Frampton down, I found, I'd feel unsatisfied. I'd want Gary to prostrate himself outside my window. I'd want him to *submit.*

I don't know. Maybe it was because I was used to having my own way at home when I was growing up—my brother was eleven years older, making me very much the baby of the family—but when I went away to college I found myself astounded, really astounded, by the number of fools I encountered. When I got to the White House, I expected things to be different. Everyone, I assumed, would be completely professional, proving just as pleasant, competent, and committed to the President's agenda as was the President himself. Instead, I discovered, the White House contained so many fools that you could look at practically anyone on the entire organization chart and wonder, quite legitimately, what he was doing there.

Human imperfection. Watching Ronald Reagan deal with it taught me how to do so myself—even when the imperfection was my own.

——*Them*——

WITHIN MONTHS of becoming a presidential speech-writer, I found myself with a couple of perplexing questions about the people on Reagan's staff. To understand my questions, you'll need a word of background.

Broadly speaking, the White House staff divided into two camps, the pragmatists and the true believers. The goal of the pragmatists was simple. They wanted to persuade Ronald Reagan to stop acting like a conservative, becoming a moderate instead. The pragmatists urged the President to scale back his tax cuts, advised him to mention his pro-life beliefs only infrequently, and implored him to tone down his anti-Soviet rhetoric. The goal of the true believers was just as simple as that of the pragmatists but directly opposed to it. They wanted the President to pursue every aspect of his conservative agenda, enacting, to use their own term, a Reagan revolution.

The pragmatists, whose numbers included the first and second chiefs of staff, James Baker and Donald Regan, dominated the White House staff. This placed us speechwriters, true believers to a man, in the position of street urchins: to survive, we had to fight. "The speechwriters were a band of brothers," Dana Rohrabacher says. "We were the musketeers. We were the Reagan underground. Our enemies were the Soviets, liberal Democrats, and the White House senior staff, although, come to think of it, I just listed them in the wrong order." In an earlier chapter, you'll recall, I suggested that you could trace the whole story of Reagan's victory in the Cold War by looking at just four speeches: the address to the British Parliament, the "evil empire" speech, the Berlin Wall address, and the address to students at Moscow State University. You already know how bitterly the pragmatists fought the Berlin Wall address. Yet they fought the first two speeches almost as bitterly—not until he published his memoir, *Eyewitness to Power*, some seventeen years after the event, did David Gergen, director of communications for much of Reagan's first term, admit that in attempting to thwart the "evil empire" speech he'd been mistaken. The pragmatists failed to oppose only the last of the four speeches, the address at Moscow State University, which the President delivered when his victory in the Cold War was already apparent. The best illustration of the hostility between the pragmatists and the speechwriters, however, lay not in the open battles the pragmatists forced upon us but in the constant harassment to which they subjected us. Consider an incident in 1985 that irks Josh Gilder and me to this day.

The occasion was the death on October 29 of Ambassador John Davis Lodge, a member of the prominent Boston family (Lodge's brother, Senator Henry Cabot Lodge Jr., was Richard Nixon's vice presidential running mate in 1960). Josh Gilder received the routine assignment of drafting a statement, praising the life of Ambassador Lodge while offering condolences to his family, of the kind that the President issued whenever a prominent American died. In one of the statement's two short paragraphs, Josh described Ambassador Lodge as a "fighter against Communism" and a "constant friend of freedom," both of which were quite true. When the statement came back from staffing, however, only one of the two encomiums remained. The person in charge of all the paperwork that went to the President, the staff secretary, whom I'll call the papermeister, had deleted the phrase "fighter against Communism." Unable to believe his eyes, Josh trooped from his office to mine to show me. For a moment or two I could hardly believe my own eyes.

"It's not as if I've written some sort of vicious attack on Gorbachev that's going to make headlines around the world," Josh said. The statement would be quoted only in the obituary notices of a few newspapers in the Northeast. "Nobody *cares*," Josh said. "So why would anyone cut this?"

Maybe, I suggested, the papermeister hadn't had a reason. Maybe he'd deleted the words simply because they hadn't felt right. "Fighting against Communism?" I said. "These days? Isn't that a trifle passé?"

"Right," Josh replied, picking up my sarcasm. "Why besmirch a man's reputation at the hour of his death by insinuating that he understood the Soviet threat?"

"Or cause his family needless embarrassment," I added, "by lumping him with that coarse vulgarian Ronald Reagan?"

Once we'd gotten started, it took Josh and me no time at all to work ourselves up into a state of truly righteous indignation. There was nothing unusual in that, of course. But this time we noticed something unusual. The papermeister had left himself open to a counterattack.

"When I challenge him on it," Josh said, suddenly delighted, "he'll have to back down. I mean, he'll *have* to. He'll recognize he was just responding to some sort of weird prevailing sense of *comme il faut*. Then he'll imagine what would happen if word of what he's done reached the President. Am I right?"

In the event, Josh was right.

Josh telephoned the papermeister and, doing his best to sound naïve, asked him to explain why he had deleted the phrase "fighter against Communism" in the first place. The papermeister replied that he considered it inappropriate. "Gee," Josh said, sounding even more naive, "Ronald Reagan has spent more than thirty years as a 'fighter against Communism.' Do you suppose the President would consider the phrase 'inappropriate'?" For a moment the papermeister was silent. Then he began to splutter—as I recall, he accused Josh of being confrontational. And then he agreed to let Josh put the phrase back in.

"Here we are," I said to Josh after that little victory, "almost six years into the administration of Ronald Reagan, the man who called the Soviet Union an 'evil empire.' But we *still* have to put up with a staff secretary sitting not fifty paces from the Oval Office itself who thinks the phrase 'fighter against Communism' is somehow 'inappropriate.' How can this *be?*"

"The big fights they start are bad enough," Josh said. "But this little stuff? This harassment? This stuff that's intended to trip us up and wear us down and undermine our morale? It *never* stops."

Josh and I ended our conversation by asking ourselves the first of my two questions. Ronald Reagan had won the presidency by standing for a vivid, well-defined set of beliefs. Why had he staffed his administration with so many people who failed to share them?

—— *Us* ——

THIS BRINGS ME to the true believers. Although within just a few days of joining the President's staff I'd begun to recognize the faults of the pragmatists—I could hardly miss the way they'd try to water down our speeches—I failed to see any shortcomings among the true believers until weeks later. I was one of the true believers myself, after all. I inhabited their camp. Yet eventually even I had to admit it, if only to myself. Examine the people who gathered around our campfire, warming them- selves in the glow of ideological commitment, and you'd find— well, you'd find quite a few odd cases. And none would be any odder than those you'd discover among that band of brothers, that troop of musketeers, the speechwriters themselves.

What did I see when I myself looked at Josh Gilder, Clark Judge, Dana Rohrabacher, and Tony Dolan, the speechwriters I knew best? Easy. Richly talented writers and thoroughly ad- mirable men. What did I see when I looked at them as the prag- matists must have looked at them? That, too, I'm afraid, was easy. Cranks. To the pragmatists, I recognized, my fellow speechwrit- ers could hardly have seemed any more out of place in the White House if they'd been actual musketeers, in broad-brimmed hats with ostrich feathers, who went around calling themselves Athos, Porthos, Aramis, and D'Artagnan.

The pragmatists understood money. They'd spent years work- ing with it. And they'd made a lot of it. As a member of one of Houston's leading law firms, for example, James Baker had spent two decades putting together deals for the oil, natural gas, and other industries, while Donald Regan had served as chairman of Merrill Lynch, one of the largest financial institutions in the na- tion. My fellow speechwriters? They could scarcely understand their own checking accounts. One had his secretary balance his checkbook for him. Another never even seemed to grasp the meaning of a balance. "If I ever bounce a check," he once told me, "I'm sure somebody will let me know."

The pragmatists had enjoyed glittering careers, moving smoothly along ever-rising arcs to positions of power and influ- ence in law, finance, and academia. My fellow speechwriters had never had careers. They'd spent their time knocking around instead. When he graduated from Harvard Business School, Clark Judge could have gone to Wall Street. Instead he'd gone to the Upper West Side, where he'd rented a walk-up apartment, then written articles on economics and political history, acted as a one-man consulting firm, and, although a Republican in one of the most left-wing neighborhoods in the nation, dabbled in New York City politics. When he graduated from Yale, Tony Dolan could have landed a job in a big corporation. Instead he'd landed gigs in Greenwich Village, where he'd appeared in restaurants and clubs as a folk singer—a *conservative* folk singer, you understand. On the album he recorded, Tony included a song he'd written, "Join the SDS," about the hard-left organization Students for a Democratic Society. "Now Dr. Spock is with them," the song went, "and that I'm gladly for. No other protest movement needs a baby doctor more." It was a wonderful song—I thought so when Tony performed it in my office, anyway. Somehow, though, "Join the SDS" failed to carry Tony's album into the Top 40.

Serious, composed, earnest, and level-headed, the pragmatists had mastered the practical realities of politics and life. My fellow speechwriters? Please.

One day a few months after accepting his job at the White House, Josh Gilder learned that his security clearance was taking an unusually long time to go through. "What do you think it could be?" Josh asked, demonstrating his bewilderment before the workings of the world. "I mean, I used to be a Democrat, but that's not against the law, is it?"

"What about your family?" I asked. "Any dubious characters in the family tree?"

Josh thought for a moment. "Come to think of it, in the 1950s my grandfather was blacklisted as a Communist."

"Your grandfather was a *Communist?*"

"I don't think he was a card-carrying Communist himself. But at one point the union he helped to found, the Transport Workers Union, cooperated pretty openly with the Communist Party. Could that be what's holding up my clearance?"

"Just maybe."

Dana Rohrabacher retained all the attributes of the college kid he'd been when he served as a student volunteer in Reagan's 1966 campaign for governor of California. Although he kept a suit and tie on a hook on the back of his door, Dana would work at his word processor wearing sandals, cutoffs, and a Hawaiian shirt. One day shortly before we were due for a speech meeting in the Oval Office, Dana pulled his suit and tie down from the back of his door, changed into them, then realized that he'd left his shoes at home. Dana began pounding up and down the marble-tiled hallways of the Old Executive Office Building, frantically looking for someone who'd lend him a pair of shoes. After waiting as long as we could, the rest of us left for the West Wing. Moments before we were to be shown into the Oval Office, Dana joined us in the anteroom. He'd borrowed a pair of wingtips. The shoes were at least four sizes too big, but that did nothing to dampen his sense of accomplishment. When the door opened, Dana flip-flopped happily in.

This brings me to an awkward point. You see, if there was one speechwriter at whom the pragmatists would have been most justified in raising their eyebrows, I was conscious—no, *acutely* conscious—that that speechwriter was me. Josh Gilder, Clark Judge, Dana Rohrabacher, and Tony Dolan may not have understood money, but they'd always paid their bills. Me? Half the reason I'd left England was to escape dunning letters from my Oxford college, which kept wanting to know when it could expect my final tuition payment. Josh, Clark, Dana, and Tony may not have pursued careers, knocking around instead. But me? Spending a year in a dank Oxford cottage writing one-half of an unreadable novel didn't even count as knocking around. Josh, Clark, Dana, and Tony may not have been the most practical of people, but before going to work at the White House they'd all mastered reality well enough to spend several years earning their livings in the private sector. Me? I'd never held even one full-time job. Far from mastering reality, I'd scarcely encountered it.

That band of brothers, that troop of musketeers. Whenever the subject of us speechwriters came up, I recognized, the pragmatists must have shaken their heads in disbelief that we'd ever been hired. Which brings me to my second question. Ronald Reagan took pride in running a businesslike and efficient administration—not long after I joined his staff, a national business magazine even ran a cover story on Reagan's management techniques. Why did he tolerate cranks like us?

—— *How to Close a Barn Door* ——

THE SPEECHWRITER most likely to understand the White House as an organization, I decided, was Clark Judge—as I've said, he'd graduated from Harvard Business School—so one night when I gave him a ride home after we'd both been working late I asked him my two questions. I even told Clark how ridiculous I'd concluded the speechwriters, including Clark himself, must look to the pragmatists. He took it better than I'd expected. "Offbeat, eccentric, naïve. I know how they see us," he said. Then we sat for a while at the curb outside his apartment building.

"Why does the White House contain two kinds of people as different as the pragmatists and the true believers?" Clark asked. "As I see it, you've got three basic explanations."

The first of Clark's three explanations was political. "When Reagan gets to the Republican convention in 1980 he's defeated all the more moderate candidates—George Bush, Howard Baker, Bob Dole, all of them," Clark said. "The conservative true believers have been with Reagan from the start. But what's he going to do with all the moderates who supported the other candidates? Just shut them out of the Republican Party? Not a chance. He has an election to win."

Instead of shutting the moderates out, Clark argued, Reagan had to gather them in. "One way he does that is by naming George Bush as his running mate. But another way is by bringing James Baker, Dick Darman, and a whole slew of other pragmatists into his campaign." The pragmatists were in the White House, in other words, because Reagan had needed their support to capture the White House in the first place.

Clark's second explanation concerned Reagan's management style. "The cry of the true believers may be 'Let Reagan be Reagan,' " Clark said. "Let Reagan be Reagan" was a catchphrase among administration conservatives. In Washington during the Reagan years, you could see the phrase on bumper stickers. "What the true believers don't understand," Clark continued, "is that Reagan *is* being Reagan."

Reagan, Clark pointed out, had spent six years as a union leader. "The members of the Screen Actors Guild were serious people. And they had serious money at stake. They wouldn't have elected Reagan to six terms as President if he hadn't been a skilled negotiator." Details of negotiations varied, of course, but fundamentally any good negotiator had to know how to take just two steps, staking out compelling positions and then cutting good deals.

"Where do the true believers come in?" Clark asked. "We help Reagan stake out his positions. And where do the pragmatists come in? They help Reagan cut his deals." The speechwriters, for example, might fashion a compelling call for tax reductions. Then pragmatists such as James Baker and Dick Darman would sit down with leaders of the House and Senate to negotiate, cutting the deal. "It's not that we're right and they're wrong or that they're right and we're wrong," Clark said. "Reagan is very deliberately using us both."

Clark's final explanation concerned Reagan himself. The chief executive, Clark maintained, was bighearted, generous, and forgiving. When true believers complained about his compromises or pragmatists sought to undermine his stands, Reagan simply forgave them both. "Nixon never forgave a subordinate a single mistake," Clark said. "Reagan? It's hard to imagine what you'd have to do to get him to fire you."

David Stockman, the director of the Office of Management and Budget, presented a case in point. Early in the administration Stockman granted a series of interviews to the journalist William Greider in which Stockman derided the very economic program he was pledged to put into effect, mocking Reagan's policies as "trickle-down" economics. When Greider's article about Stockman appeared in the December 1981 issue of *The Atlantic Monthly*, it created an outcry. Nearly all Reagan's advisers urged the President to fire Stockman. ("Deaver and I wanted him fired," Ed Meese now says, "and you can be pretty sure the first lady wanted him fired, too.") Yet instead of firing Stockman, Reagan merely gave him a talking-to.

"Think back to Hollywood," Clark said. "Anybody who's read any show business memoirs knows Reagan ran into plenty of alcoholics, drug addicts, and people with sexual tastes that the folks back home in Dixon, Illinois, would have had trouble believing. But Reagan saw how capable Hollywood professionals were even so. He saw them write scripts and design costumes and act and direct, providing affordable entertainment for millions."

As in the motion picture industry, Clark argued, so in politics. "The movie business and politics are both messy and collaborative," he said. "Big egos, fights over speeches, people taking wild fliers like the one the Bush staff took when they hired you— Reagan is willing to tolerate all of that as long as the job gets done. If you want to accomplish anything important, he knows, you have to forgive people an awful lot."

AFTER MY TALK with Clark, I began looking at the pragmatists in a new light. Not, of course, that I stopped fighting them. We speechwriters would have done Ronald Reagan a disservice if we'd permitted our work to become as watered-down as the pragmatists would have liked. Yet now I recognized that the pragmatists had as much right to their place in the administration as we true believers had to ours. The pragmatists knew how to plan schedules, cut deals, and manage the federal government. They mastered large bodies of information of the kind we speechwriters couldn't be bothered to learn. A couple of the pragmatists in the West Wing actually kept copies of the federal budget in their offices, referring to the budget just as often as I referred to my dictionary. Their turn of mind wasn't my turn of mind. But I had to grant that it had its uses. I still saw a lot of the pragmatists as small-minded, petty, and incapable of under-standing the rightness of the President's conservative agenda. Yet instead of railing against them, I did my best to overlook their shortcomings. Actually, that's only half true. Railing against the pragmatists was too much fun to give up. But I did do my best to overlook their shortcomings.

If you want to accomplish anything important, Reagan un-derstood, you have to forgive people an awful lot. Some months after my talk with Clark, an incident took place that comes to mind whenever I consider Reagan's capacity for overlooking the faults of those with whom he worked.

To appreciate the incident, you need to understand that in every White House the congressional liaison is considered sus-pect. This is only fitting. His job is to persuade Congress to do what the President wants. But when he leaves the White House he's likely to become a lobbyist—serving as congressional liaison doesn't lead to a lot of other positions—so he's always tempted to please the barons of Capitol Hill by persuading the President to do what Congress wants instead.

This dynamic came into play when Reagan's own congres-sional liaison conducted a briefing on tax measures one day in the Cabinet Room. When the liaison reached a measure the President especially wanted enacted, he hung his head.

"I'm sorry about this, Mr. President," the liaison said, "but I've spoken to Senator Dole [Bob Dole, then Senate majority leader], and he says the horse has already left the barn on this one."

The President looked disappointed. "You've spoken to Bob Dole?"

"Yes, sir," the liaison replied.

"And he says the horse has already left the barn?"

"Yes, Mr. President. It's unfortunate, I know. But the horse has already left the barn."

The President studied the congressional liaison for a moment. Then, as the liaison continued the briefing, the President wordlessly stood and left the room. Several minutes later, he returned.

"I just got off the telephone with Bob Dole," the President said. "The horse is back *in* the barn."

You didn't have to be a presidential historian to imagine how other chief executives would have handled that incident. Richard Nixon would have submitted the congressional liaison to an icy glare, then instructed his chief of staff to teach the man a lesson. Lyndon Johnson would have subjected the congressional liaison to a stream of profanities on the spot. Ronald Reagan? He used the incident to get a laugh—and then he let it go. Reagan proved just as bighearted, generous, and forgiving as Clark had said.

—— *Heart One Place, Head Another* ——

IF RONALD REAGAN taught me to treat people I might otherwise have looked down on, such as the pragmatists, with respect, he also gave me a couple of pointers about how to behave when the fault lay not with others but with myself. Let me introduce this subject by describing another incident at college.

Every evening at Dartmouth I'd get together in Thayer Hall, the dining hall, with the same rowdy buddies. We'd discuss sports, argue over politics, talk about women, and play practical jokes with our food, dumping salt in someone's clam chowder when his head was turned or stealing someone else's hamburger when he went back to the food line for more French fries. One evening I snatched a baked potato from the plate of my friend

Tim, pulled back his collar, and dropped the potato down his shirt. Tim didn't think it was funny. What surprised me was that nobody else at the table thought it was funny, either. First everyone shouted that I was out of line. Then, when I refused to apologize—in one of my Type A phases, I argued that no one should have to say he was sorry for a simple practical joke—everyone waved his napkin at me. I walked off in a huff.

My dinner buddies were my closest friends. Now they'd reached a unanimous decision that I'd broken some sort of unspoken rule. Even after I'd calmed down it took me awhile to figure out just what that unspoken rule had been. I hadn't just messed up Tim's food, I finally saw. I'd assaulted his person. Rude, crude, and unshaven they may have been, but my college buddies retained a finely calibrated sense of their dignity. Why hadn't I seen that unspoken rule for myself? What was I? Some sort of incipient psychopath? I apologized to Tim, but not before brooding for days.

I describe this incident because it provides a neat illustration of two kinds of shortcomings, the outright mistake—in this case, dropping the potato down Tim's shirt—and the blind spot or limitation—in this case, my inability to see an unspoken rule that everybody else considered obvious. Ronald Reagan presented an example of each kind of shortcoming himself.

Reagan's shortcoming of the first kind, his outright mistake, was his misjudgment—or rather his series of misjudgments—in the Iran-contra scandal. Since even today Iran-contra remains the principal blot on Reagan's record, the scandal is worth taking a moment to consider.

The allegations against Reagan were twofold. In selling arms to Iran, according to the first, Reagan attempted to purchase the freedom of American hostages. Such an arms-for-hostages deal violated American policy—and common sense, for that matter—by rewarding terrorists for taking Americans hostage in the first place. The second allegation concerned the proceeds of the arms sales. Lieutenant Colonel Oliver North, a midlevel member of the National Security Council staff, sent money from the arms sales to the contras, the anti-Communist guerillas who were waging an insurgency against the Sandinista regime in Nicaragua. This action, it was widely assumed, violated the Boland Amendment, which was intended to place strict limits on American involvement in Nicaragua.* North acted with the knowledge of John Poindexter, then the national security adviser. According to the second allegation, the President himself either ordered this diversion of funds or granted it his tacit approval.

The second allegation is easily dismissed. It's untrue. After spending more than seven years and $35 million investigating Iran-contra, independent counsel Lawrence Walsh found "no credible evidence" whatsoever that Reagan knew about the diversion of funds until the scandal itself broke. John Poindexter and Ollie North acted on their own.

Yet the first allegation, that Reagan approved an arms-for-hostages deal, contained some truth. The President never intended a bald swap. But the arms deal he approved included a dubious element all the same. Consider the meeting that took place on January 7, 1986, to discuss whether the President should sign a "finding," a document that would give the arms sales legal standing. Those joining the President in the Oval Office included CIA director William Casey, national security adviser John Poindexter, secretary of defense Caspar Weinberger, secretary of state George Shultz, and attorney general Edwin Meese.

---

* In the event, no violation of the Boland Amendment was ever proven in court.

Presenting the terms of the deal, which involved selling defensive armaments to elements in Iran, Casey and Poindexter gave the President four reasons to support it. First, they argued, the arms sales would open lines of communication between the United States and moderates in Iran. (One of the moderates Casey and Poindexter identified was Ali Akbar Hashemi Rafsanjani, who later became President of Iran.) Casey and Poindexter's next three reasons each involved goals the Iranian moderates had agreed to do their best to achieve: bringing the war between Iran and Iraq to a rapid conclusion, thereby denying the Soviets, who had massed troops on the border of Iran, any opportunity to intervene; reducing Iranian support for international terrorism; and persuading Hezbollah, the terrorist group operating in Lebanon, to free the seven American hostages it was holding.

Shultz and Weinberger objected. "They said it would look too much like arms for hostages," Ed Meese now says. "That wasn't what it was—we'd be asking the Iranians to use their influence with a third party, Hezbollah, to get our people back—but they argued it was still too close to that." Meese himself granted the weight of argument on both sides. "I said, 'Mr. President, it's a fifty-one to forty-nine percent decision,'" Meese explains. Then he sided with Casey and Poindexter, urging the President to support the deal. Reagan did so.

When the scandal broke—by way of a leak to *Al Shiraa*, a newspaper in the Middle East, possibly by the very Iranians to whom the arms sale was supposed to endear the administration—the President quickly took a bad situation and made it even worse.

On November 13, he delivered a speech that contained a number of inaccuracies, including the claim that all the weapons and spare parts the United States had supplied to Iran "could easily fit into a single cargo plane," which was flatly untrue. Then, on November 19, he held a press conference in which he once again misstated certain facts, denying, for example, any participation in the arms shipments on the part of Israel, which had instead been directly involved.

In neither the speech nor the press conference did Reagan make any attempt to deceive the public. The speech, which was written largely by Patrick Buchanan, then director of communications, was based on bad information supplied by national security adviser John Poindexter—bad information that the administration later corrected. The President's statements during the press conference were likewise based on bad information, in this case briefing papers prepared by both John Poindexter and former national security adviser Bud McFarlane, who later testified that the material he provided "was not a full and completely accurate account." Even the President's denial of Israeli involvement in the arms shipments represented only an innocent mistake. "There was so much discussion before the press conference of what was and wasn't classified," Ed Meese explains, "that the President got confused. When he was asked if another nation was involved, he gave the wrong answer. But there was no attempt to deceive." Far from it. The White House press office issued a correction less than half an hour after the press conference ended. Yet in the speech and press conference alike, the President himself went before the public and stated untruths.

As if these first mistakes were somehow insufficient, the President compounded them by spending the next four months insisting he'd never traded arms for hostages in the first place. He hadn't—not exactly, anyway, as we have seen—but it looked to millions of Americans as though he had, and in refusing to admit that an arm's-length transaction is still a transaction, the President made himself seem intransigent.

Now on the scale of recent presidential misdeeds—Richard Nixon obstructed justice, for example, while Bill Clinton lied under oath—Ronald Reagan's offenses scarcely register. One bad decision, one botched speech, one bungled press conference, and one stubborn streak. That was all. Yet Iran-contra prompted me for the only time during my six years in the White House to wonder what had gotten into the President. The one bad decision? The President, I concluded, should have listened more closely to Ed Meese. In the meeting on January 7, 1986, you'll recall, Meese advised the President that the weight of argument tipped narrowly in favor of the arms shipments. Yet Meese supported only tentative overtures to Iran. If they proved successful, Meese held, the overtures could be developed. If not, they should be dropped. Reagan knew something Meese didn't. Israel had *already* made arms shipments to Iran, in effect engaging in the very overtures Meese thought the United States was now about to make. Since the Israeli overtures had produced few results, Meese's own argument suggested the President should forbid the arms deal, not support it.

The botched speech and bungled press conference? Reagan knew that the arms deal over which John Poindexter had presided had become a fiasco. Before agreeing to either a speech or press conference, the President should therefore have brought in another national security expert to review Poindexter's actions, establishing the facts—and then the President should have mastered those facts himself. The stubborn streak? "My heart and my best intentions still tell me [it's] . . . true [that I did not trade arms for hostages]," Reagan finally admitted in a March 4, 1987, address from the Oval Office, "but the facts and the evidence tell me it is not." Heart one place, head another. That's what life is like for all of us from time to time, and polls indicated that the American people accepted this explanation. Why couldn't Reagan have offered the explanation months earlier?

The President had made a series of mistakes, and he knew it. How, I wondered, would Reagan respond? Over the months during which Congress and investigators probed Iran-contra, the President addressed the scandal in a number of decisions and speeches, but his actions all came down to three simple steps. He held himself accountable, opening his November 19 press conference with a statement in which he said, "[T]he responsibility for the decision and the operation is mine and mine alone." He did everything he could to rectify his errors, dismissing Ollie North, accepting John Pondexter's resignation,* and instructing everyone in the administration to cooperate with the many investigations that were soon under way. And then? He got *on* with things, continuing to show up at the Oval Office every morning at nine o'clock. For a number of weeks he felt low—everyone in the White House could tell that just by looking at him—but he never indulged in long, brooding conversations about his troubles as did Richard Nixon and Bill Clinton, devoting himself instead to his work, and soon enough even his usual high spirits returned. When the scandal broke in late 1986, some predicted Reagan would end 1987 a lame duck, his administration adrift. Instead he ended 1987 by hosting Mikhail Gorbachev at the White House for the signing of the Intermediate Range Nuclear Forces, or INF Treaty, an event so significant that one historian, Niall Ferguson, argues that it ended the Cold War.

---

* Although former national security adviser John Poindexter and Lt. Col. Oliver North were both convicted of a number of charges arising from the Iran-contra scandal, both saw their convictions overturned on appeal.

Human imperfection. When you make a mistake, Reagan showed me, you should hold yourself accountable, do whatever you can to set matters right, and then, instead of brooding, get *on* with things.

—— *The Old Man* ——

R EAGAN'S SHORTCOMING of the second kind? His blind spot or limitation? Despite his warmth and geniality, he could have been a better father. "We all wanted more of him," Ron Reagan says. "That was especially true of Maureen and Mike [Reagan's children by his first marriage]. But none of us could get enough of him."

As I was working on this book, Michael and Ron each told me a story that captured his relationship with Reagan. Mike's took place one day as he and his father were driving to the ranch in Malibu. Then an adolescent, Mike asked for a bigger allowance. "I was getting five bucks a week, and he really made me earn it," Mike says. "I'd paint fences out at the ranch and do things like that." How did Reagan reply? "He told me how much he paid in taxes—that ninety-one cents out of every dollar went to the government. If I was looking for a bigger allowance, the best thing for me to do was figure out how to get a new President in place that would give us a tax cut."

Ron's story took place one Sunday morning as the family was getting ready for church. "I was only twelve, but I'd decided I was an atheist," Ron says. "My father came in to fetch me, and I was just sitting there in my PJs. I told him I wouldn't be going. He looked kind of crestfallen. He tried briefly to reason with me, quickly realized that this was something that was going to take a lot more time, and then said, 'Okay, it's your call, but, gee, I'm kind of sad about this.' "

Reagan got in touch with Donn Moomaw, the minister at Bel Air Presbyterian, the church the Reagans attended. "My father invited him up to the house to convince me to go back to church. It didn't work," says Ron, an atheist to this day. "Donn talked to me years later and said, 'I was so embarrassed. What was I supposed to say to you?' "

In neither story did Reagan appear in any way callous or abusive. Yet in both he missed opportunities so obvious that even an outsider like me could see them. Why didn't Reagan ask Mike why he needed a bigger allowance in the first place? Mike might have replied by telling his father about girls he wanted to date or trips he wanted to take, permitting Reagan in turn to tell Mike what his own life had been like when he was Mike's age. Father and son might have been able to open up to each other. They might have been able to connect. Why didn't Reagan ask Ron how he'd decided he was an atheist? Ron might have replied by telling his father about books he'd read or conversations he'd had, permitting Reagan in turn to explain what his faith meant to him. Once again, father and son might have been able to connect. Why hadn't Reagan *seen* that?

"He was a terrific father up until adolescence," Ron says. "But when adolescence hits, it's no longer going out in the backyard and throwing the football around. Instead you want to learn adult stuff and have adult conversations. You want to be brought into the world of adulthood by a parent. That's when the relationship [with my father] started to sort of slip away. An adolescent is an adolescent—a big pain in the ass. That made him very uncomfortable."

Strictly speaking, Reagan's relationship with his children was none of my business. He was the President, I was one of his speechwriters, and that was that. But you couldn't work at the White House without noticing the contradiction or paradox. Reagan loved his children. That was clear. I worked several times with Maureen, an activist in the Republican Party, on events at which she and her father appeared together, and in each appearance the President all but glowed with affection for her. Yet the Reagan children seldom visited the White House, Camp David, or Rancho del Cielo, while stories describing family tensions appeared constantly in the press.

Had Reagan proven distant, I wondered, because his own father, an alcoholic, had provided him with such a poor role model? Had his myopia—until he began wearing glasses as an adolescent, Reagan had been unable to see clearly for more than a few feet— somehow caused his world to close in on him, depriving him of the opportunity to learn how to establish intimate relationships? Or had Reagan proven remote from his children simply because of the era in which he had grown up? "When my father was born the *Titanic* was in dry dock and World War I hadn't started yet," Ron Reagan says. "We were still essentially in the Victorian era. Emoting all over the place just wasn't something you learned."

Ronald Reagan was such a transparently good person, I finally concluded, that if he'd known how to become a better father he'd have done so. The trouble was, he didn't know how. He doesn't seem to have had any more idea than his children themselves what made it so difficult for him to draw close to them—and although if you read their memoirs you'll encounter plenty of spirited speculation about Reagan's remoteness, you'll also get the feeling that the Reagan children were never quite able to figure it out. Every man has his blind spots. This was one of his.

After Reagan left the White House, his children seemed to find ways to accept their father's limitations.

"There was no malice in him," Ron says. "He was the best father he knew how to be." Maureen, who died in 2001, grew so devoted to her father that she served as a fundraiser for Reagan's alma mater, Eureka College, sat on the board of the Alzheimer's Association, and made frequent trips from her home in Sacramento to that of her father and stepmother in Bel Air to care for her father as he deteriorated, attempting to keep his mind alert, for example, by helping him solve jigsaw puzzles. "I consider it a good day when I get several smiles and laughter," she once wrote. "There's nothing like the sound of his laughter." Patti, now reconciled with her parents, helps to care for her father while serving as her mother's chief support, striving to keep up Mrs. Reagan's spirits. Critical of her father in earlier books, she praises him in her 1995 book, *Angels Don't Die: My Father's Gift of Faith*. "The world," she writes, ". . . should know that [Ronald Reagan] passed along to his daughter a deep, reliant faith that God's love never wavers." Mike sees now that his father sought over the years to reach out to him. "That was why he made a point of taking us [children] to the ranch," Mike says. "The ranch was always his territory. He took us there so we could get to know his heart—just know the love and warmth that was inside this man."

Human imperfection. This may be a book of lessons I learned from Ronald Reagan, but here I learned a lesson from his children. They proved big-hearted, generous, and forgiving enough to love their father as he was.

—— *Children Overhead* ——

GARY AND his Peter Frampton album? I should have lightened up. All Gary ever wanted to do was blast some music for a while before he settled down to study, not start a dorm war. Tim and the baked potato? Tim deserved an immediate apology, of course. And although my failure to see beforehand that I'd be breaking an unspoken rule when I dropped the potato down Tim's shirt *still* bothers me, I shouldn't have been surprised to learn that I have my limitations. We all see some things but miss others.

I behave better now than I did in college. The lessons I learned watching Reagan handle the White House staff have helped me in every organization in which I've ever worked, teaching me to treat everyone, even those I find difficult, with respect. And the lessons I learned by considering Reagan's own mistakes and limitations have helped me in life itself. When I make a mistake, I try to own up to it, make amends, and get on with things. My limitations? Although I think of myself as close to my children, I feel my limitations most acutely as a father even so. There's just so much I don't *know*.

How much time should I insist the children devote to their studies? How much should I encourage them to devote to sports instead? One son has announced that he wants to drop soccer for football. Which would be best for him? Since he'll do all his work on computers when he grows up, another son has announced, his sloppy handwriting doesn't matter. He's right, of course, but he's also wrong. What should I tell him?

Working on this book one weekend I heard a thudding sound overhead. Running outdoors, I found every child but the baby clambering across the roof. In the middle of a game of Cops and Robbers, I learned once I got the children back on the ground, one of my sons had hidden behind a trellis, discovered the trellis was sturdy enough to climb, and then shared this discovery with his big sister and little brothers. I gave the children such a harsh talking-to that Edita decided no further punishment was needed. I couldn't decide whether I'd been too lax with the children beforehand—if I'd been a good father, I thought, they'd have known they weren't allowed on the roof without my having to tell them so—or too severe with them afterward. But by the time I got back to my desk I'd decided the incident represented a hopeful sign. Children who can figure out how to climb onto a roof without a ladder will probably find their way in life, despite their father's limitations.

--- *Ten* ---

THE LIFEGUARD
VS. KARL MARX

You Matter

AFTER THE PRESIDENTIAL ELECTION of 2000, you
may recall, the *New York Times* printed a map of the nation
in which the counties that George W. Bush carried appeared in
red, those that Al Gore carried in blue. The two Americas, the red
and the blue. Bush's support lay primarily in the nation's interior,
the great heartland comprising the South, the plains, and the
Rocky Mountain states. Gore's support lay chiefly on the coasts.
Yet there was more to the two Americas than their location, as I'd
learned in writing a book about American politics not long before.
Characterized by small towns and medium-sized cities, red
America was the land of people who subscribed to *Reader's Digest*,
attended church on Sundays, and took traditional morality for
granted. When you looked at paintings by Norman Rockwell,

what you saw was red America. Dominated by the nation's largest conurbations, including the vast megalopolis that sprawls from Boston to Washington, D.C., blue America was by contrast the land of people who subscribed to the *New York Times*, read up on yoga and Sigmund Freud, and took the sexual revolution for granted. When you watched *Seinfeld*, what you saw was blue America. Red America, in a word, was square; blue America, cool.

When I worked at the White House I knew both Americas from the inside. I'd grown up in the red America (upstate New York has changed over the years, but during my boyhood it was as red as any place in the nation), then moved to the blue America, taking up residence in Washington, D.C. Like an immigrant who resists certain customs of his new land, I refused to take on the politics of blue America, which, as the election of 2000 so neatly demonstrated, tended toward the Democratic. But I made my living like a lot of people in blue America, manipulating words and ideas, and although I was a conservative, not a liberal, I got just as much of a kick out of feeling I was participating in cultural and political life—that I was, in short, cool—as any denizen of the Upper West Side of Manhattan. Soon, though, I discovered I had a problem. It was a problem a lot of immigrants face. I found myself with mixed feelings about those I'd left behind.

My feelings centered on one person, my father. As things go between fathers and sons, we'd always had a pretty smooth relationship, and we remained close even after I left home. But my father was such a perfectly typical product of red America, so decent, plodding, and square, so lacking in urbanity or sophistication. I worked with my mind. He worked with his hands, a graphics designer at IBM, where he produced drawings of machinery. Never a manager, he remained a working stiff throughout his career. I graduated from Dartmouth and Oxford. He never attended college. I flourished in Washington, D.C. He always disliked big cities, once telling me how, as an aspiring painter at the Art Students League, he'd felt out of place in New York City. I read the *New York Times*, the *Washington Post*, the *Wall Street Journal*, and more than a dozen magazines. He read *Reader's Digest* and the *Binghamton Evening Press*, and that was that. It made me uncomfortable to do so, but I just couldn't seem to help it. I looked down on him.

Ronald Reagan got me over it.

To explain how, I need to borrow a phrase from the philosopher Sidney Hook, showing you Reagan as a "hero in history."

—— *Our Hero* ——

"WHEN HISTORY is taught at all nowadays," George F. Will wrote not long ago, "often it is taught as the unfolding of inevitabilities—of vast, impersonal forces. The role of contingency in history is disparaged, so students are inoculated against . . . the notion . . . that history can be turned in its course by . . . individuals." Working for Ronald Reagan amounted to a graduate course in just the opposite, the ability of a single man to change the entire world.

Yes, I know. Margaret Thatcher may believe the fortieth chief executive won the Cold War—Lady Thatcher, as I've noted, has often said that "Ronald Reagan won the Cold War without firing a shot"—but her view is hardly universal. Well, then. If Ronald Reagan didn't win the Cold War, how *did* the conflict end? There are only a couple of other explanations.

One holds that the Soviet Union simply collapsed, like a

poorly designed astrodome after a storm dumped tons of snow on the roof. What placed the fatal stresses on the USSR, the explanation goes, were economic stagnation, imperial overreach—that is, an empire that had grown so big the Soviets could no longer afford it—and the rise of a generation that failed to share the Communist faith of its parents and grandparents. Reagan? Don't be silly. He had nothing to do with it.

Or had he?

The Soviet Union certainly did suffer from economic stagnation. But the Soviet economy had been growing feebly since at least the early 1970s. What changed during the 1980s wasn't so much the economy of the USSR as the economy of the United States, which, as we have seen, responded to the policies of Ronald Reagan by growing dramatically. By the time Reagan left office, the American output of goods and services had expanded by an amount nearly equal to the entire economy of what was then West Germany. The only way the Soviets could have expanded their economy by that amount would have been by annexing West Germany itself. If the Soviets finally decided they'd had it with the creaking, backward economic contraption that Stalin, Khrushchev, and Brezhnev had given them, in other words, they did so because they'd caught a glimpse of the sleek new beauty that Ronald Reagan had given us.

Imperial overreach? True enough, the Soviets found themselves stuck with an empire they could no longer afford. But you can hardly blame them. By rebuilding our military, Ronald Reagan had forced the Soviets to spend more on theirs. By arming the contras in Nicaragua and the mujahedeen in Afghanistan, he had compelled the Soviets and their proxies to engage in long, expensive wars of attrition merely to cling to territory they had already come to think of as their own. By delaying the construction of the Soviet natural gas pipeline and working with oil-producing countries, including Saudi Arabia, to hold down the price of oil, a principal Soviet export, he had deprived the Soviets of vital sources of hard currency. Costs up, revenues down. The Soviet case of imperial overreach came courtesy of Ronald Reagan.

Did a new generation of Russians refuse to place its faith in the Communism of their forebears? Evidently. But why? In part, surely, because of the transformation young Russians saw taking place in the United States.

During the 1970s, the United States looked like a nation in decline, just about as Karl Marx would have predicted. Even President Carter contended the country had seen better days. "The symptoms of . . . [a] crisis in the American spirit are all around us," Carter said in an address from the Oval Office on July 15, 1979. Then, on January 20, 1981, Ronald Reagan took office. "The crisis we are facing today," he said in his first inaugural address, ". . . [requires] our willingness to believe in ourselves and to believe in our capacity to perform great deeds . . . And after all, why shouldn't we believe that? We are Americans." As polls indicated, the American people responded with a renewed sense of patriotism and self-confidence. "Morning in America," the campaign slogan for Reagan's 1984 reelection campaign, was widely derided in the media, as presidential campaign slogans tend to be. Yet it captured the mood of the nation. If you'd like proof, take a look at Reagan's 1984 victory. He carried forty-nine out of fifty states, receiving more votes than any other candidate for President in history.

Morning in America? As the children of the Soviet apparat would have recognized, that wasn't in Marx's game plan. Ronald Reagan made Communism look a lot less like the wave of the future and a lot more like another misbegotten nineteenth-century ideology, such as syndicalism or anarchism, that was destined for the ash-heap of history.

The other explanation holds that the Cold War ended because a spiritual and cultural revolution swept Eastern Europe. Led by Pope John Paul II, Lech Walesa, and Vaclav Havel, the velvet revolution of 1989 toppled one Eastern European regime after another until, in a reversal of the domino theory, the USSR itself came crashing down.

The explanation has a certain amount to be said for it. "The Pope would have done just what he did, Reagan or no," Richard Allen, Reagan's first national security adviser, explains. Walesa, Allen believes, "may have been a bit more cautious absent Reagan." But then again, maybe not. "As for Havel, he found great solace in Reagan's presence. . . . [but] he was already committed [before Reagan became President]." It appears certain, in other words, that the Pope, Walesa, and Havel would have done all they could to foment a revolution in Eastern Europe even if Reagan had never left Hollywood.

What appears a lot less certain, as I tried to suggest in the exchange with Christopher Hitchens I described in an earlier chapter, is that the Pope, Walesa, and Havel would have succeeded. Reagan gave the revolution in Eastern Europe very substantial support. He provided funding and equipment to Solidarity in Poland, for example, both covertly, by way of the CIA, and overtly, by way of the AFL-CIO, whose assistance to Solidarity the administration coordinated. Still more important, Reagan let the Soviets know that as events in Eastern Europe unfolded he expected them to keep their hands off. "We grew concerned [in spring 1981]," Richard Allen says, describing just one incident, "about Soviet troop movements in and around Poland." To prevent the Soviets from crushing the Solidarity movement in Poland just as they had crushed the 1968 Prague Spring, on April 3, Reagan, still in the hospital after the attempt on his life, sent a blunt message to Brezhnev, threatening, as Reagan later wrote, "the harshest possible economic sanctions." Did Reagan's action forestall a Soviet invasion? If a definitive answer exists, it lies buried in the archives of the Politburo. Yet soon after Brezhnev received Reagan's letter, the Soviet forces stood down. The revolution of 1989 might not have proven as soft as velvet if Ronald Reagan hadn't spent eight years beforehand proving as hard as steel.

Ronald Reagan won the Cold War. The Soviet Union collapsed. Pope John Paul II, Lech Walesa, and Vaclav Havel led a revolution. Feel free to choose any explanation for the end of the Cold War that you like. But if you settle on either of the final two, sooner or later, I think, you'll notice something peculiar. Even though the explanation you've chosen directs your attention not to decisions made in Washington, D.C. but to events on the Eurasian landmass, it will make a lot more sense if you bear Ronald Reagan in mind.

ON THE NIGHT in 1976 when Ronald Reagan lost the Republican presidential nomination to Gerald Ford, Michael Reagan was seated with his father in a Kansas City hotel not far from the convention hall. "This was the first time in my life I'd ever seen my father lose," Mike says. "So I asked him, 'Dad, what's going through your mind?' "

What disappointed him most, Reagan replied, was that now he'd be unable to face the leader of the Soviet Union at the negotiating table. "My dad said to me, 'Michael, I wanted to sit down with Brezhnev. I was going to allow him to choose the place. I'd even allow him to choose the room, choose the table, and choose the chairs. I wanted to sit there and listen to him tell me, the American President, representing the American people, everything the United States was going to have to give up just to get along with the Soviets. Michael, I wanted to listen to him for fifteen or twenty minutes. Then I was going to get up from my chair very slowly while he was talking, and I was going to walk around to the other side of the table. And then I was going to lean over and whisper in his ear, '*Nyet*.' "

Nyet. What other politician of the day so yearned to utter that syllable to the General Secretary of the Soviet Union? Jimmy Carter? Hardly. Not long after taking office in 1977, Carter announced that under his leadership the United States was at last free of its "inordinate fear of communism." Carter's fellow Democrat, Walter Mondale? Once again, hardly. Running against Reagan for President in 1984, Mondale denounced the fortieth President for his hard line with the Soviets. Even among Republicans, what other politician wanted to say *nyet* to Leonid Brezhnev? George H. W. Bush? Howard Baker? Bob Dole? Yet again, hardly. Running against Reagan in the 1980 Republican presidential primaries, each placed himself to Reagan's left.

"The great man or woman in history," the philosopher Sidney Hook argues in his book, *The Hero in History*,

> is someone of whom we can say on the basis of the available evidence that if they had not lived when they did, or acted as they did, the history of their countries and of the world, to the extent that they are intertwined, would have been profoundly different. Their presence, in other words,

must have made a substantial difference with respect to some event or movement deemed important by those who attribute historical greatness to them.

Does Reagan fit the description? He does indeed. No one else would have done what he did. And what he did changed the world. But you needn't take my word for it. "He was an authentic person and a great person," Mikhail Gorbachev said in a recent interview. "If someone else had been in his place, I don't know if what happened would have happened."

There you have his principal adversary all but admitting it: Ronald Reagan was a "hero in history."

—— *The Lifeguard* ——

ALTHOUGH REAGAN always thought of himself as a good actor—"the worst thing you could do to him," Lyn Nofziger says, "was to say he was only a B actor. He'd start ticking off the names of all the famous people he'd starred with"—no one I know ever heard Reagan boast about his success in politics, let alone claim to be a hero. Yet Reagan always seems to have believed that he could influence the outcome of the Cold War. "It was as if he thought he had been put on earth to crush the Soviet Union," says Jeffrey Hart, an English professor at Dartmouth who took a leave of absence to write speeches for Reagan during his 1968 presidential campaign. "You got the feeling that when he shaved every morning he'd look at himself in the mirror and ask what he could do to hurt the Soviets that day."

Now when you think about it, this is just a little bit odd. Take a look at the intellectual movements that dominated the nine-

240

teenth and twentieth centuries. Darwin argued that we're the crea-
tures of our genes, Freud that we're the products of our sexual de-
sires, Marx that we're the subject of a kind of cosmic dialectic, an
interplay of actions and reactions that are themselves subject to certain predetermined patterns or rules. Individuals? They scarcely
matter. What matters, just as George F. Will wrote, are unfolding
inevitabilities and vast, impersonal forces. Where could Ronald
Reagan ever have gotten the idea that one man could stand up to
history? Josh Gilder and I spent a lot of time wondering about that.

"If you want to understand the guy," Josh would say as he and
I sat talking in my office, "you've got to remember that as a boy
Reagan was only a couple of generations removed from the
American frontier. Laying down the first crops, building the
barns and farmhouses, putting in the railway lines—a lot of that
stuff was still a living memory." Although pioneers took up residence in Illinois during the first few decades of the nineteenth
century—Lincoln's family moved from Kentucky to Illinois in
1830, for example—it wasn't until four or five decades later that
the region took on a fully settled character, with framed houses
and barns instead of crude structures of sod and logs, with well-
established towns, and with a fully developed system of roads and
railways. "Individuals don't matter?" Josh would say. "Try telling
that to people whose grandparents busted up the sod with their
own teams of mules."

The frontier experience wasn't the only epoch Reagan would
have encountered as a living memory. Delivering a speech I'd
written, as I told Josh during one of our sessions, Reagan had
stopped when he'd reached a quotation from Lincoln. Then he'd
ad-libbed a couple of lines, telling the young people in his audience that when he was their age he'd seen Civil War veterans
march each year in Armistice Day parades. More than many con-

flicts, the Civil War had turned on individual efforts. Lincoln himself had devised Northern strategy, then seen to it that his generals pursued it, often over their bitter objections, while Grant himself had devised an array of new military techniques for bringing the North's superior wealth and numbers to bear upon the South. "Individuals can't affect history?" I'd say. "Try telling that to men who fought under Lincoln and Grant."

Settling the land and preserving the Union. Reagan believed individuals capable of remarkable feats because he'd imbibed aspects of the American experience that proved it.

And because his own experiences had confirmed it. Working summers from 1927 to 1932 as a lifeguard at Lowell Park, where a stretch of the Rock River had been designated for swimming, Reagan had saved from drowning no fewer than seventy-seven people. "All that matters is impersonal forces?" Josh would say. "Right. There were seventy-seven people walking around northern Illinois who wouldn't have been there if it hadn't been for Reagan—and Reagan knew it."

"I sort of like the metaphor here," Josh would continue. "History as a river. Marx says all we can do is surrender ourselves to the currents. Reagan says, 'Oh yeah? Well, I can handle the currents. I've done it before.' "

Often enough this would bring Josh and me to the matter of Ronald Reagan's religious beliefs. As we have seen, Reagan seldom discussed his faith, but everyone in the White House sensed it all the same. I'd raise the subject gingerly—in those days Josh himself was a skeptic, not a believer—noting that even an agnostic had to admit Christianity implied a lot about the importance of individuals. An individual Himself, Jesus died to redeem not abstractions, but men and women.

"I'll grant you that the rules in the Bible are pretty clear,"

Josh would say. "God doesn't relate to the world by dealing with impersonal forces. He doesn't even deal with governments very often—except to shake them up, the way He did when He sent plagues to Pharaoh. What He deals with is individuals— Abraham, Moses, the prodigal son, the woman at the well. God created history for individuals, not the other way around."

As handsome and talented as he was, Josh and I agreed, Reagan might have found himself drawn to the Nietzschean belief that the only individuals with the power to stand up to history are the superior few, the *übermenschen* or supermen. "But Nietzsche just isn't what's in his head," Josh would say. "What's in his head is simple Christian belief as mediated by the democratic ideals of small towns in the Midwest. Reagan doesn't believe he's a superman. He believes *everybody's* a superman—that all individuals have a kind of supreme inherent dignity."

"The way he looks at it," Josh would continue, "he just happens to be the one whose background and talents made him President. But he doesn't consider himself one bit better than anybody else. And with him it's not just an intellectual notion. It's real. It really is. It's so much a part of him that you can see it in the way he treats people every day."

You could indeed.

JOURNAL ENTRY, NOVEMBER 2002:

Stopping by the office this afternoon to pick up my mail, I ran into Martin and Annelise Anderson. {Colleagues of mine at the Hoover Institution, the Andersons worked on Reagan's presidential campaigns, then served in his administration.} They both looked beat. When I asked what had brought them into the office on a Saturday, they replied that they'd been working nonstop for days now in order to meet the deadline on their newest book, a collection of Reagan's letters {in 2000, the Andersons and a third

Hoover fellow, Kiron Skinner, published a collection of Reagan's radio talks and speeches titled Reagan, In His Own Hand*}. "Every time we think we've got all his letters pulled together," Martin said, "more of them pop up."*

When Martin asked me to guess how many letters Reagan composed while he was President, I thought back to a conversation I'd had in the White House with Anne Higgins, the director of presidential correspondence. Out of the thousands of letters that arrived for the President each week, Anne had told me, she'd choose ten or twelve to give him. Each weekend Reagan would take the letters Anne selected to Camp David, read them—and almost always answer every one. A dozen letters a week, fifty-two weeks in a year, eight years in the presidency. "If what Anne used to say is true, then I'd put the number at about five thousand," I said. "But to be honest, I always wondered if Anne was exaggerating. I mean, the man was President. How could he have found the time to handle that much correspondence?"

Martin shook his head wearily. "Five thousand? Not a chance. It looks as though the final number could be as high as nine thousand."

Collecting the letters had proven so arduous, Martin explained, because Reagan had sent so many to the kinds of people who don't keep careful files. "He had his regular correspondents—people like Bill Buckley and Walter Annenberg—but a huge number of letters went to people Reagan didn't even know," Martin said. "He was just writing to ordinary Americans."

B EFORE WORKING for Ronald Reagan, Martin Anderson worked for Richard Nixon. Traveling with Nixon during one of Nixon's many campaigns, Martin listened in one day as Nixon gave an interview to a reporter on the campaign plane.

Putting his notepad away after concluding the interview, the reporter asked Nixon what part of politics he found hardest.

"Some people think the hardest part of politics is making the decisions on policy issues," Nixon replied. "That's not it. Some people think it's all the traveling and speechmaking. That's not it. Some people think it's raising money. That's not it. The hardest part of politics is getting the bastards to vote for you."

You could imagine Ronald Reagan's chatting with a reporter after an interview. You couldn't imagine his calling the American people "the bastards," even in jest. The thought would never have framed itself in his mind. Reagan saw no difference between important people, such as Presidents, and unimportant people, such as ordinary voters. People, to Reagan, were people. "My father treated his secretary the same way he'd have treated a foreign potentate," Ron Reagan says. "I never in my life saw him condescend to anybody."

In the White House, you'd see one instance after another of Reagan's complete evenhandedness. Walking from one part of the complex to another, Reagan might stop to talk to anyone who said hello, paying as much attention to the comments of a typist as to those of a member of the senior staff. In the speechwriting shop, the secretaries got in the habit of checking the President's schedule for events that would bring him from the West Wing to our own building, the Old Executive Office Building, then gathering on the big balcony that lay just off our front office to wave to the President as he crossed West Executive Avenue. The President always took the time to stop, wave back, and shout up a few words of greeting. (One spring day I joined the secretaries on the balcony myself. After the President waved and shouted hello, he spotted a mound of melting snow. He scooped up a couple of handfuls, patted them together, then tossed a snowball up at us. The Secret Service agents glared, willing us to refrain from returning fire.)

Then there was the story of Frances Green, a tale that was told and retold, becoming the Reagan administration's very own folk legend. But as folk legends go, this one had an unusually good provenance. It was true.

A little old lady who lived in Daly City, California, Frances Green contributed a few dollars a year to the Republican Party. As a token of its gratitude, the GOP apparently sent Mrs. Green a fake invitation, suitable for framing but not for anything else, of the kind still common in the direct mail industry. Failing to read the fine print, Mrs. Green concluded that she'd been asked to meet the President. She spent four days crossing the country by train, then presented herself at 1600 Pennsylvania Avenue. When word of Mrs. Green somehow reached him, the President invited her in, sat her down in the Oval Office, apologized for the delay in admitting her, and chatted with her about California, showing Frances Green the same courtesy he'd have shown Elizabeth II.

Yet the incident I always considered the best illustration of Reagan's regard for ordinary individuals took place not in the White House but in a North Carolina parking lot. "It was during the 1976 primary fight," says Dana Rohrabacher, who then worked on the Reagan campaign as an assistant press secretary. "We were getting ready for a rally in this gigantic parking lot at a shopping mall. I was in the staging area behind the podium, and a lady called me over to the side and said, 'I've got a group of blind kids here. Since they can't see him, I was wondering if you could have Governor Reagan come over and tell them hello.' "

Dana passed the request along to Mike Deaver, and Reagan, who was standing nearby, overheard. "He said he'd do it, but he didn't want any photographers," Dana explains. "Can you imagine that? He was in the middle of a presidential campaign, and the press would have gone wild for a photo of him with a group of blind kids. But Reagan wanted this to be between him and the kids."

Deaver came up with a plan. When the speech ended, Deaver told Dana, he'd begin walking Reagan back to the campaign bus. Concluding that the candidate was about to leave for the next event, all the reporters and photographers would hurry back to their own buses. And then, when the press had cleared out, Deaver would double back with Reagan, returning the candidate to the area behind the podium, where Reagan would meet the blind children.

"It worked," Dana says. "The press guys all went back to their buses, and I brought the lady with the blind kids back behind the podium. There were six or seven kids, real sweet little kids about eight or nine or ten years old. Since there was a lot of background noise—you know how it is after a speech, with a crowd breaking up—Reagan bent down, close to the kids, to talk to them. But somehow I could see him thinking that that wasn't enough. So after the kids had asked him a couple of questions, he said, 'Well, now I have a question for you. Would you like to touch my face so you can get a better understanding of how I look?' The kids all smiled and said yes, so Reagan just leaned over into them, and one by one these little kids began moving their fingers over his face to see what he looked like.

"The only picture of that scene is the picture in my mind," Dana says. "But I can still see those little kids, touching Ronald Reagan's face and smiling these really big smiles."

"The Declaration of Independence," G.K. Chesterton writes, "dogmatically bases all rights on the fact that God created all men equal; and it is right [to do so]. . . . There is no basis for democ-

247

racy except in a dogma about the divine origin of man." Although in nearly every way you could ever imagine, in other words, we humans are not equal but *un*equal—some rich and some poor, some bright and some dull, some healthy and some sick—in one way we enjoy perfect equality all the same. Did the fortieth chief executive ever read Chesterton? I can't say. Yet Ronald Reagan demonstrated an implicit belief in the sacred and equal importance of all men as children of God.

This brings me back to my father.

—— *The Boiler Tender* ——

JOURNAL ENTRY, MAY 2002:

I still can't get over it. My father only even shook hands with Ronald Reagan twice in his life, once at a Christmas reception at the White House, and once when the President spoke in Endicott, New York, just across the Susquehanna River from our home in Vestal, and I was able to arrange for my parents to have their picture taken with him. But as I toured the Reagan ranch today, I kept thinking of Dad.

When Marilyn, my guide, pulled up her Ford Explorer after the long climb up the mountain, I opened the door, got a whiff of the horses—the tang of manure was unmistakable—and thought of the way Dad had loved horses ever since he was a boy hanging around the stables of his Uncle Jimmy, the family rogue, who made his living in harness racing. Then, when I saw the firewood piled next to the front door of the ranch house—it's old firewood now, the last pile Ronald Reagan split himself— I once again thought of Dad. How many hours had Dad spent splitting and stacking firewood himself every autumn? Inside the ranch house, I looked over the bookshelves—and yet again, I thought of Dad. I counted a couple of classic novels by Horatio Alger Jr.; several dozen works of pop-

ular modern fiction, including Larry McMurtry's Lonesome Dove; *one book after another about horses; and a couple of pictorial histories of the old West. If Dad had for some reason found himself at the Reagan ranch with a few days on his hands, he'd have been perfectly happy to choose three or four of Reagan's books, sit down, and start reading.*

After showing me the ranch house, Marilyn took me up the hill to the big steel shed that served as a garage and tack room. At the back of the shed stood Ronald Reagan's workbench, a simple wooden structure beneath a pegboard. The pegboard held saws, hammers, screwdrivers, and other tools, each hung neatly in its place. I could scarcely believe it. Reagan's workbench was almost identical to Dad's. You could have dropped Ronald Reagan into our garage or my father into Ronald Reagan's steel shed, and, repairing a chair or fixing a drawer, either man would have been able to pull down the tools he needed almost without having to look.

Rounding out the tour, Marilyn drove along one of Reagan's riding trails. When she came to a knoll, she pulled over and stopped. Under a stand of live oak lay Reagan's animal graveyard, the place he'd buried his horses, his favorite bull, and his dogs, including Rex, the King Charles spaniel that Mrs. Reagan had named after Rex Scouten, the White House usher. Over each grave, Reagan had laid a stone on which he had chiseled the animal's name. Although Reagan's graveyard was much bigger, it reminded me of the graveyard that Dad had put in the woods behind our house for our pets. When my guinea pig died, I remembered, we'd spent quite a while trying to decide what to chisel on the headstone. Since the full name, Supercalifragilisticexpialidocious, the nonsense word from Mary Poppins, *wouldn't fit, we'd finally settled on "Super."*

As Marilyn and I drove back down the mountain, I had an odd daydream, picturing Dad working on the ranch right alongside Dennis LeBlanc and Barney Barnett, the two men Reagan employed, and Reagan himself. Brushing down the horses, mucking out the stables, splitting firewood, digging postholes, pruning trees, working for hours at a time while saying barely a word, Dad would have fit right in. He and Reagan were the same kind of men.

IN THE MATTER of my father, Ronald Reagan presented me with a couple of problems. If Reagan accorded everyone the same perfect courtesy, to name the first problem, then who was I to look down on anyone? The second problem was even worse. The more I saw of Reagan, the more I recognized that he and my father had a lot in common. What sense, I felt forced to ask myself, did it make to look up to only one of them?

Although, as my journal indicates, I found new parallels between Reagan and my father when I visited the Reagan ranch last year, I recognized plenty of similarities between them while I was still at the White House. Both had been born in the second decade of the twentieth century, Reagan in 1911, my father in 1919. Both had grown up in modest circumstances, Reagan the son of a shoe salesman, my father the son of a man who spent much of his career hauling goods around the floor of a Sears, Roebuck warehouse. Both possessed a simple, straightforward Christian piety that they'd learned from their mothers. Both had found themselves drawn to the arts, Reagan to acting, my father to painting. And both had remained citizens all their lives of what we would now call the red America: decent, patriotic, low-brow in their tastes, men who loved a good joke, enjoyed the outdoors, and found themselves most at ease when working with their hands or engaging in hard physical labor.

All right, I thought. Maybe they *did* have a lot in common. But Reagan had gone places and done things. He'd become a star. My father? All he'd become was a working stiff. Was there anything my father had ever done that Reagan hadn't? Just one. During World War II, he'd seen action. Whereas Reagan had served in a movie unit, my father had served aboard a Coast Guard cutter that was assigned to antisubmarine duty. His cutter had sailed in both the Atlantic and the Pacific, where my father had found himself present at the Battle of Okinawa. Yet since my father hardly ever talked about the war—he'd mentioned that it hadn't been any fun to watch the kamikazes attack the fleet at Okinawa, but that was about all—I seldom gave his experiences any thought. Ronald Reagan changed that.

At the 1984 ceremonies in France marking the fortieth anniversary of the Normandy invasion, Reagan made clear his reverence for fighting men. Although ordinarily composed when he spoke, the President had tears in his eyes when he stood at the summit of the cliff at Pointe du Hoc that American Rangers had had to scale and then, the survivors seated before him, delivered the beautiful lines that Peggy Noonan had composed: "These are the boys of Pointe du Hoc. These are the men who took the cliffs." Yet during Reagan's state visit to France two years earlier, an incident took place that I found even more striking. I heard the story from Evan Galbraith, at the time our ambassador to France.

Midway through a meeting at the ambassador's residence, deputy chief of staff Michael Deaver stepped in to say that he'd just learned the French intended to award the President a decoration.

Reagan looked puzzled. "What decoration?"

"I think it's the *Croix de Guerre*," Deaver replied.

"The *Croix de Guerre*?" Reagan said, his face clouding. "But that's for bravery. All I ever did was fly a desk. We'd better get this straightened out right away. I couldn't possibly accept the *Croix de Guerre*."

Fearing a diplomatic incident, Galbraith himself stepped out to talk to the French. "It's not the *Croix de Guerre* they'll be giving you after all," Galbraith said, returning to the meeting. "It's the *Légion d'Honneur*."

"What for?" Reagan asked, puzzled once again.

"Statesmanship," Galbraith replied.

Reagan relaxed. Then he tightened the knot in his tie and grinned. "Statesmanship?" he said. "I can play that role."

Although amusing, the incident gave me pause. If Reagan felt such profound respect for men who had been exposed to the dangers of battle, I thought, maybe I'd better reconsider my father's own wartime experience. Although my father had never received any medal, for some three years he had served as a boiler tender aboard the *Roger B. Taney*. My father would spend hours at a time in the boiler room, where he regulated the oil that was piped into the boilers to keep them burning and the superheated steam that was piped out of the boilers to power the ship's engines. Deep in the ship, next to the boilers. That would have been a bad location if a torpedo or kamikaze had ever hit—and during the Battle of Okinawa, I learned when I did some research, the *Taney* had come under repeated attack by kamikazes, sounding general quarters 119 times in just 45 days. In my father's place, I knew, I'd have spent every moment frightened of finding myself scalded or trapped. As it turned out, that would have been the wrong fear. The pipes in the boiler room were wrapped with asbestos, and the vibrations from the ship's engines, which were located in an adjoining compartment, kept the air in the boiler room full of asbestos particles. Thirty years after the war, my father lost part of a lung to asbestosis.

A working stiff? If Ronald Reagan had known about my father's wartime service when they met, I realized, then the

President of the United States would have given my father his crispest salute.

——*The Long Handshake*——

RONALD REAGAN'S RESPECT for each individual taught me a new way of looking at individuals myself. Although a "hero in history," Reagan understood that greatness is possible on any scale.

When my father graduated from high school in the midst of the Depression, for example, the only job he could find was digging ditches with the Johnson City, New York, department of public works. I never heard him speak a word of complaint about it. During World War II he did his duty and then, when he came home, he got married and started a family, taking a series of jobs he didn't much enjoy because he had a wife and children to support. He raised my brother and me, taught us what was important in life, and made it possible for us to attend college, receiving the kind of education of which he himself had scarcely been able to dream. And when he died, of conditions exacerbated by asbestosis, becoming, in effect, a delayed casualty of World War II, he left us with an example of decency, goodness, and strength.

No one would say the world is different because my father lived, of course, but *his* world is different because he lived. My brother and I are leading good lives. Our children, my father's grandchildren, are coming along just fine. On the scale of life as it came to him, my father was a hero in history himself.

Now that I've reached middle age, I've found myself surrendering dreams I've entertained since adolescence. I know I'll never become President, never start a company that will make me fabulously rich, never act on Broadway, and never compose the greatest American novel since *Huckleberry Finn*. I'll never have much effect on the world, in other words. But I can still have an effect on my own world, and I find this consolation so much more specific and concrete—so much more *real*—that I've begun to like it even better than the dreams. I can perform my work as well as I can, hoping to provide my readers with sturdy, enjoyable prose. I can remain loyal to my friends and a loving husband to my wife. And I can remind myself every day that from the moment of their birth until the very end of time, I'm the only father my children will ever have.

A scene comes to mind. It's the Christmas reception at which my father first shook hands with Ronald Reagan. As the Marine Corps band plays Christmas carols in the background, my father, my mother, and I stand behind a velvet rope in the East Room of the White House with a couple of hundred other people as the President and Mrs. Reagan work their way down the rope, shaking hands. Just as Reagan clasps my father's hand in both of his, someone behind my father asks Reagan a question—something about taxes, as best I can recall—and Reagan, still clasping my father's hand, answers the question at some length, talking for at least half a minute before he finally finishes, drops my father's hand, and continues down the rope. I can still see that moment: the crowd, my mother, Mrs. Reagan, the President, and my father, gazing at Reagan with a look of reverence and awe, scarcely able to believe he was shaking hands with the President of the United States. That look on my father's face, I used to think, was funny—another manifestation of his squareness. Now what I see in that look is the reverence and awe I feel for my father himself. If I give my children reason to feel half as much for me, I will have done enough.

EPILOGUE

IN THE SPRING OF 1989 I stopped by the suite of offices that had just been set up for former President Reagan and his staff. As he stood to greet me, Reagan had the same twinkle and shine in his eyes and the same knowing nod that he had possessed in the White House. "Just doing a little writing," he said, gesturing to a pad of paper on his desk. "Now that I'm out of office, I have time to get back to writing my speeches myself."

After a moment of small talk, Reagan frowned and asked if I'd seen the morning newspaper. I had, noting over breakfast that the *Los Angeles Times* referred to Reagan in two front-page stories. "SAW RISK OF IMPEACHMENT, MEESE SAYS," one headline read, while the other stated, " 'STAR WARS' WAS OVERSOLD, CHENEY SAYS."

"I just don't understand it," the former President said.

"Neither do I, Mr. President."

"How can a *judge* decide the outcome of a sporting event?"

It took me a moment to realize that Reagan wasn't talking about his administration. He was talking about the America's Cup. A judge in New York had just awarded the Cup to the boat from New Zealand, even though the American boat had put in a faster time. "SAN DIEGO LOSES AMERICA'S CUP," the headline stated, "CONNER'S USE OF CATAMARAN RULED TO BE VIOLATION OF GOVERNING DEED."

"Well," Reagan said, the twinkle returning to his eye, "at least it wasn't a judge *I* appointed."

When I left, I felt disappointed at first that the former President hadn't even mentioned world events, let alone imparted any secrets or insights of historical moment. I'd had a moment with the man who won the Cold War, and all I'd come away with was some talk about a boat race.

But by the time I was back in traffic on the Santa Monica Freeway, I recognized that Reagan had given me a good example of the wisdom and simplicity of spirit that I'd always cherished in him. You see, I was still having trouble getting used to the idea that I no longer worked at the White House myself. The excitement, my buddies in the speechwriting shop, the opportunity to observe the President—I missed them all a lot more than I'd expected. And even though I knew it was ludicrous, I couldn't quite shake the thought that when I'd written the Berlin Wall address, I'd peaked. Would the rest of my life amount to nothing but a long anticlimax?

Reagan snapped me out of it. For eight years he had been the most powerful man in the world. Then he had set it all down and gone back to being just as ordinary an American as a former President can be. When he picked up a newspaper, he turned to the sports page.

There are plenty of satisfactions, the former chief executive helped me to see, in being an ordinary American.

I now begin the journey that will lead me into the sunset of my life. I know that for America there will always be a bright dawn ahead.

—*Ronald Reagan, letter to the American people,*
November 5, 1994

Reagan knew what Alzheimer's disease would do to him. He'd watched the disease claim the mind of his own mother. "Back then it was called 'senility,' " Michael Reagan says, "but Nelle had Alzheimer's, too." On Sunday afternoons Nelle Reagan would drive from Los Angeles to Olive View Hospital in Simi Valley, where she worked as a volunteer. As she grew more forgetful, Reagan himself made the drive one day, stopping at all the shops and gas stations along the way to leave pictures of his mother at each. "On the back was Dad's name and phone number," Mike says, "so if Nelle forgot who she was or where she was going, the store managers and gas station attendants would be able to call Dad to come get her." Yet even as Alzheimer's disease in turn claimed his mind, Reagan conducted himself with dignity.

Watching the former President in the mid-1990s, John Barletta, a Secret Service agent assigned to Rancho del Cielo, grew concerned. At first Reagan proved incapable of controlling El Alamein, the Arabian that had for many years been his favorite mount. "As he was riding El Alamein home one day," Barletta says, "we had a problem. I had to get off my horse and grab his horse." Barletta replaced El Alamein with a gentler mount. Soon, however, the former President began having trouble with his horse's tack. "He was just making mistakes that a rookie would make," Barletta says. "Knowing what a great rider he was, that worried me. Even saddling, cinching up, putting the bridle on, things like that—he was having trouble with them. He would stop and stare at the bridle, and you could tell he was trying to figure out which way it went."

At last Barletta decided he could no longer permit the former President to mount up. "I went to Mrs. Reagan and I said, 'Mrs. Reagan, he can't ride anymore. He's making too many mistakes.'" Afraid that if she broke the news to her husband he might suppose she was being overprotective, Mrs. Reagan asked Barletta to tell Reagan instead. "She said, 'He'll take it much better coming from you.'

"It was after lunch or dinner that I went down [from the Secret Service post, which sat on an overlook above the ranch house]—Mrs. Reagan told him I was coming down. And I told him. I said, 'Mr. President, you know all the problems we had on the trail today. I really don't think . . . I really don't think you should ride anymore.' I was breaking up. And he got up. And he put his hands on my shoulders. And he said, 'It's okay, John. I know.'"

One final glimpse:

"There's a particular Secret Service agent, an African-American man," Ron Reagan says. "This guy is a big, burly guy who looked like he played football. When my father was still able to get up and on his feet and sort of be walked to the living room and put in a chair, this guy would be on one side of him and another agent on the other. My father would often not want to let go of his hand."

Once when the agent had helped Reagan to his chair, the former President seemed to want to express his thanks. "He couldn't verbalize anything at that point," Ron says. "But he looked at the agent. "Then he took his hand and kissed it."

A few bright stars in the night sky. To orient the telescope I aimed it at them, marveling as always at their colors—orange Aldebaran, yellow Capella, blue Vega. . . . As often happens, I was struck by the fact that all these things, unimaginably big or small or hot or cold as they may be, really are out there.
—*Timothy Ferris,* Seeing in the Dark

Yes, I know. It seems odd to close with a quotation about stars. Yet how often do we find ourselves thinking about historic figures the same way we think about celestial bodies—as distant and impersonal and somehow unreal? Ronald Reagan was born in a town that still stands. He liked to tell jokes and watch old movies. If you visit his ranch, you can see his favorite riding boots, standing in his tiny closet, the leather still supple. For that matter, you can see his burro, Wendy, a stubborn, comical-looking creature he seems to have bought simply because he liked the idea of having her around. Wendy is old now; her coat, once gray, has turned almost entirely white, and she waddles across her hillside corral slowly and stiffly. But she'll still let you approach to scratch her behind her enormous, floppy ears. She's used to being scratched behind the ears. That's where Ronald Reagan scratched her.

Soon enough, the fortieth President of the United States will slip away, departing the narrow confines of the present to take his place in the annals of the nation. From that hour on we will need to remind ourselves that he was no mere force or idea or abstraction. We can aspire to his virtues because he was one of us. Ronald Reagan was a man.

ACKNOWLEDGMENTS

I AM PROFOUNDLY grateful to everyone who took the time to talk with me about the fortieth President, discussing his place in history while revealing, often enough, his place in their hearts. Each appears in the text, but I must make special mention of three. Ever since we found ourselves trying to figure out how to write speeches together, Clark Judge and Josh Gilder have remained among my closest friends, and that's getting to be quite a while now. Their encouragement, insights into Ronald Reagan, and generosity with their time and recollections—all proved invaluable. And who'd have thought we'd have as much fun talking about the Gipper all these years later as we did in the White House itself? William F. Buckley, Jr., initiated the peculiar set of events that led to my getting a job in the White House in the first place; proved a friend, mentor, and exemplar throughout my years as a speechwriter, as he has ever since; and not only permitted me to pester him with questions as I was working on this book but persuaded others to do so as well. I believe in eternal life in part because that's how long it will take me to repay him.

Writing this book required me to gallivant around the country, place hundreds of telephone calls, and get myself some extra clerical help, and I wish to thank the John M. Olin Foundation and its executive director, Dr. James Piereson, for enabling me to do so without going into debt.

In the Office of Ronald Reagan, chief of staff Joanne Hildebrand Drake somehow spared time to assist this project in countless ways, while at the Reagan Library, archivist Gregory Cumming proved indefatigable, happily bringing out box after box of speeches I hadn't seen in almost two decades. At the Young America's Foundation, which now owns Rancho del Cielo, executive director Floyd Brown cheerfully made it possible for me to visit the ranch and curator Marilyn Fisher served as an enjoyable, patient, and informative guide. At the Baltimore Maritime Museum, director John Kellett and curator Paul Cora generously confirmed aspects of life aboard the Coast Guard cutter *Roger B. Taney.*

My assistant, Lydia Anderson, demonstrated ferocity in her pursuit of research materials but serenity in her dealings with me. My old colleague, Misty Church, who, as a researcher in the Reagan White House, used to straighten me out on the facts, straightened me out on a couple of points once again. One friend, Doug Carey, took time from his duties as a newly minted Foreign Service Officer to help me wade through economic data, while another, Arnold Beichman, three weeks shy of his ninetieth birthday, rummaged around at midnight to find me a quotation from Sidney Hook.

Acknowledgments

I am indebted to the journalist Lou Cannon for producing the monumental and engrossing book, *President Reagan: The Role of a Lifetime,* which refreshed my memory of certain events during the Reagan years, and to the historian Douglas Brinkley, whose 1999 commentary for National Public Radio, "A Clarion Call for Freedom," describes what took place inside the limousine the day President Reagan drove to the Berlin Wall. And I reserve a particular expression of indebtedness for the director of the Hoover Institution, John Raisian. Without his friendship and encouragement, this book would have proven impossible.

My agent, the inimitable Richard Pine, shaped and enlivened the project. My publisher, Judith Regan, one of the wonders of the modern world, made a couple of critical adjustments, gave the project her okay, and then kept her eye on it. My editor, Calvert Morgan—well, what is there to say? Maxwell Perkins, move over. And my unseen friend, managing editor Cassie Jones, brought the project home.

Which brings me to the six people who put up with the most while I was composing this volume. To my wife, Edita, and our children, Edita María, Pedro, Nicolás, Andrés, and Isabel: You choose the colors, and I'll paint the bedrooms.